HOW TO GROW MARIJUANA

~

MOVE FROM BEGINNER TO ADVANCED GROWER OF THIS MEDICINAL HERB

By Evan Hutchkins

Table of Contents

Introduction	1
Understanding The Stains	3
Anatomy of the Plant	10
Getting The Light Right	15
Getting The Soil Right	26
Atmospheric Conditions	36
Controlling Temperature	44
Watering Systems	48
From Growing to Harvesting	54
How to Increase Your Yield	70
Genetic Selection	75
Indoor vs Outdoor Growing	82
Troubleshooting Guide	85
Common mistakes to avoid	96
Grow Room on a Budget	101
Conclusion	115

ISBN-13: 978-1986839112
ISBN-10: 1986839117

Introduction

Over the last few years the law has been evolving in regard to growing marijuana at home. This means that there are many states which will now allow you to legally do so.
For clarity, marijuana is made from the dried leaves and flowers of the cannabis plant. It is a form of cannabis but the least potent form. The plant also secretes a gum which can be dried and smoked; this is referred to as hashish. The most potent form is hash oil which is generated from the hashish gum. Again, this is smoked.

Cannabis is a tenacious plant. It can grow virtually anywhere providing the climate is tropical and warm. Of course, if this is not true of where you live you'll need to use this guide to grow your plants inside; this is an increasingly popular choice.

The first thing to understand is that the law is complicated! Not all states allow you to grow marijuana, those that do restrict your growth to 6 plants, which may not be enough.

However, if you're a medically approved grower this amount is higher. For instance, in California you're allowed 8 ounces of cured cannabis.

If you're a registered marijuana patient, you can utilize 100 square feet to grow your plants; this is enough for approximately 20 plants. However, if you're a patient looking after to up to 5 others you can use 500 square feet of land!
It is permissible to grow marijuana in these states:

- California
- New Mexico
- Oregon
- Colorado
- Washington
- Nevada
- Hawaii
- Michigan
- Massachusetts
- Arizonia
- Alaska
- Maine
- District of Columbia

But you will need to check the law in each state; they are likely to be strictly enforced.

Sadly, this is why so many marijuana growers choose to keep their project a secret. The amount of plants you are allowed to grow in most states is not enough to ensure continual production. Read on to discover how you can grow and get the most out of your cannabis plants.

Understanding The Stains

There are three main types of marijuana, indica, sativa and hybrid. You've probably come across them in shops or in conversation. The common perception is that each type offers a slightly different experience.

- **Indica** – This is the more relaxing version of this plant. It will help you to relax and unwind, whether before bed or just chilling for the evening.

- **Sativa** – This is the opposite! It is thought to boost your energy levels, allowing you to take on physical challenges which might otherwise seem impossible. This is good if you want to be the life and soul of the party!

- **Hybrid** – As its name suggests, this is a mixture of the first two types. You won't be quite sure what you're going to get until you take it. It will depend on what the parent plants were!

Unfortunately, research suggests that these categories are misleading. In fact, the chemical make-up of these plants is unlikely to consistently provide the indicated results. Of course, there is much more research which can be done.
What is true is that the two plants, sativa and indica do have very different characteristics.

Sativa

The plant is tall with narrow leaves. It is well suited for warmer climates and, because of its long flowering cycles it will take longer to reach the harvesting stage.

Indica

These plants are much shorter and have fatter leaves. This is a distinct enough difference to make it obvious for most growers. As this has a much shorter flowering cycle it will provide quicker harvests and can cope better with colder climates.

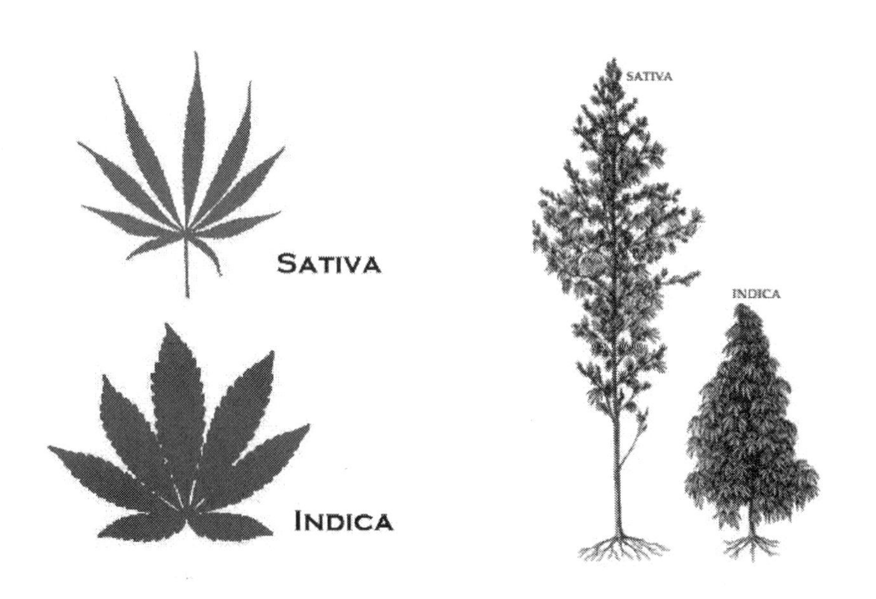

As marijuana becomes more commercialized more emphasis is being placed on which type of marijuana is being offered or desired. But not all sativa's invigorate and not all indica's will relax. It is essential to understand the wide variety of strains available to make sure you choose the right one for your needs.

The most popular strains of Indica are:

- **Bubba Kush -** Cross the bubba with the kush and you get this magical flavor. It's known to be sweet with a hashy taste. It's said to numb the mind and then spread slowly through your body. It will give you the munchies!

- **Granddaddy Purple -** This is an indoor plant grown specifically for its medicinal benefits. It originates from northern California and can be easily trained to create a large crop. In fact, this plant needs plenty of water but is very tolerant of lower temperatures. This strain will actually give you a bit of a buzz, making you feel ready to take on the world!

- **White Rhino -** This strain can be grown well in almost any environment, making it one of the best strains for a beginner. It will also provide the perfect way of slowing down, an essential requirement when dealing with many medical issues.

- **Northern Lights -** There are many versions of this strain because it can be grown in just 6 weeks. It's sweet and spicy and will relax every bone in your body. It's also very easy to grow inside!

- **Blueberry -** Unsurprisingly this strain has a taste of blueberries. It gives a long lasting buzz and will relax you, without making you feel sleepy. However, it is worth noting that this strain grows best outdoors.

While popular strains of sativa are:

- **Sour Diesel -** Imagine mixing sour lemon with the smell of diesel and you'll know what this strain is like. It's a strong odor! It's also worth noting that this plant grows 6 ft. tall. It has been said to be effective at alleviating chronic depression and is very invigorating.

- **Super Silver Haze -** This strain originated in Holland but has emerged as a firm favourite on the circuit. It gives you an enormous high which can be felt in every aspect of your body.

- **Green Crack -** Snoop Dogg is said to have given this strain its name as it will create a craving in you for more. It is excellent as medicinal marijuana because it provides a gentle, yet happy high.

- **Casey Jones -** Cross a trainwreck with Thai and some sour diesel and you'll get the Casey Jones strain. It's sweet and heady as well as fast acting. Interesting it is half sweet and half sour. It will give your mental energy a boost but only in creative ways, not logical ones.

You can opt for hybrid marijuana strains which are proving to be very popular:

- **OG Kush -** This originates in Los Angeles and is rumoured to be the most powerful hybrid available; its effects are similar to that of an espresso.

- **Chronic -** Under 600 watt lights you'll be able to get a yield of 600g per square meter! It also has a sweet and spicy taste but the effects are slower at moving through your body. It's a blend of indica and sativa giving it a gentle but forceful bite.

- **Blue Dream -** This is another strain from Northern California. It is actually a mixture of haze and blueberry. The flower is purple and blue, hence the name. When using this strain, you'll firstly feel focused and then gradually your whole body will relax. It's worth noting this is a difficult strain to grow successfully.

- **White Widow -** You'll find the white window an extremely common flavor in Amsterdam. You'll notice the flavor is a cross between flowers and fruit. It will give you a gentle high and warm your body at the same time. It is best grown inside and needs a little TLC (tender loving care) to create the perfect variant of this strain.

- **Trainwreck -** This strain has a unique smell; once you've experienced it you'll always recognize it. It is likely to have originated in Northern California. You'll be alert and very clear headed after taking this strain, helping you to resolve complicated issues which your muscles relax. It's perfect for after exercise to aid recovery.

- **Lemon Skunk -** This is a skunk crossing; two powerful lemon plants were merged to create this one. Unsurprisingly it has a citrus taste and will provide you with the opportunity to reflect while heightening your sensory perception.

- **Cinderella 99 -** This is not the greatest yielding strain but it is a popular one. You'll feel a little giddy as you reach a high and start giggling uncontrollably! This is a great strain to use to create your own marijuana plant.

- **Bubble Gum -** This is a long flowering plant, taking as much as 53 – 63 days. It was created in Holland and is known to be a slow starter. However, once it starts to grow it is a solid plant needing little maintenance. It is another coffee shop favourite.

- **White Russian -** This strain will give you a good high and is actually surprisingly easy to grow. It excels when grown indoors and can produce as much as 450g per square meter. It is sweet with a little skunk overtone and will generally make you feel more alert, although you might experience a few hallucinations.

It is important to note that the name of the strain is chosen by the plant grower. This will allow you to identify it and request it once you know which one suits you best. The name should reflect the color, taste and smell of the plant and where it originates from.

You can develop your own strain by mixing plants of different strains and even sub-mixing them! To do this you'll need to grow plants first and learn to identify their characteristics. This will allow you to select the right characteristics to keep in your own strain.

Fortunately, this is covered in detail later in the book.

Anatomy of the Plant

There are four parts of the plant a grower should be able to identify and understand the purpose of other than the leaves, stem, and roots. These are the cola, calyx, trichome, and pistil.

Cola - The cola is a section of cannabis flowers at the end of a branch. The main cola forms at the top of the plant and is where the largest collection of female flowers bloom. Smaller colas bloom lower down on the plant, but there is generally one main one at the top. The number and size of colas are directly related to genetics and growing techniques.

Calyx - The calyx is a collection of small leaves that form in a spiral where the flower meets the stem. It is part of the bud and also part of the cola. To put another way, the calyx holds the buds together. Calyxes contain high concentrations of the glands that secrete THC. These glands are called trichomes.

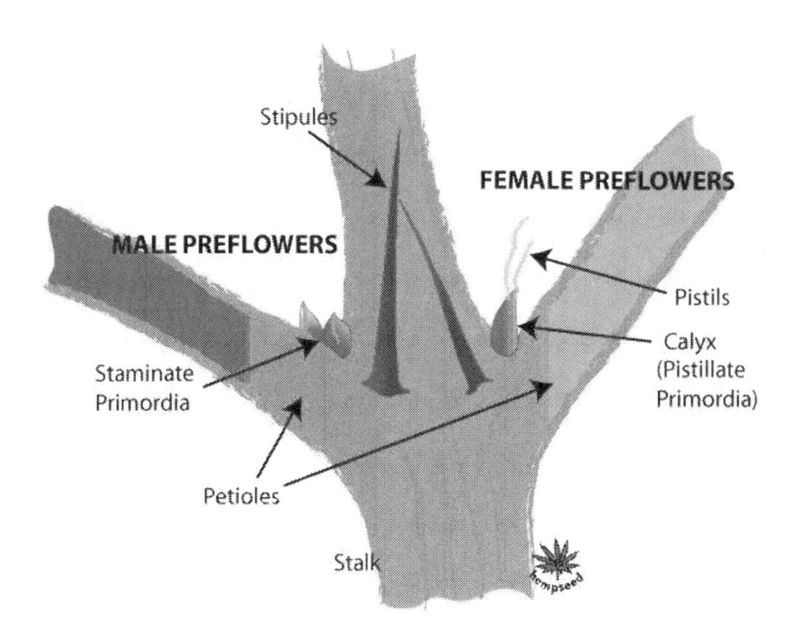

Pistil - The pistils are small hairs that grow out of the calyxes. Over the course of the growth cycle, the pistils change color from white to yellow to orange, red, and eventually brown. They're important for the plant's reproduction but don't have much of an impact on taste or medicinal effects of the cannabis.

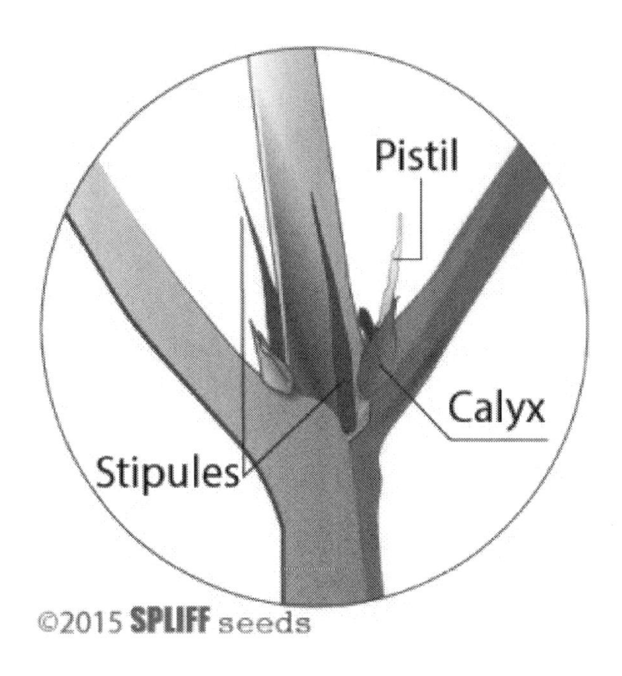

Trichome - The crystal resin you'll notice on cannabis buds (flower) is secreted through trichomes. Trichomes are mushroom shaped, translucent glands that originally developed to protect the plant from outside threats. Today, the oils they secrete- TCH and CBD- make medicinal cannabis as useful as it is in offering physical and mental relief to users.

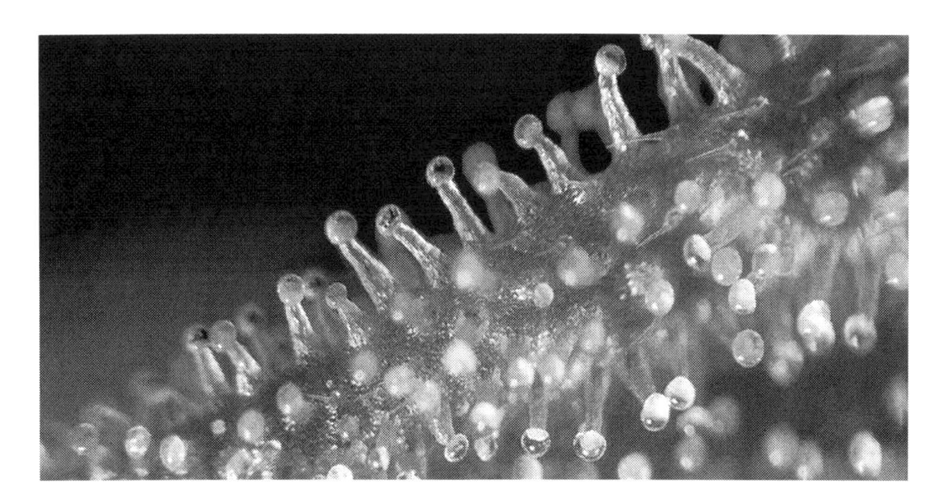

Sex of the Plant

Whether smoked or ingested, cannabis comes from the budding flowers of female plants after they have been dried and cured. Male plants will not yield flowers and cannot be used for medicinal cannabis. Until the plants begin to flower and white pistil hairs grow or pollen sacs develop, you won't be able to tell if the plant is male or female. This should take about six weeks, when the vegetative stage starts to transition into a pre-flowering stage.

As soon as you can tell the sex of the plants, remove the males so they don't pollinate the females. If a female plant gets pollinated, it will not yield as many buds or buds of the same intensity.

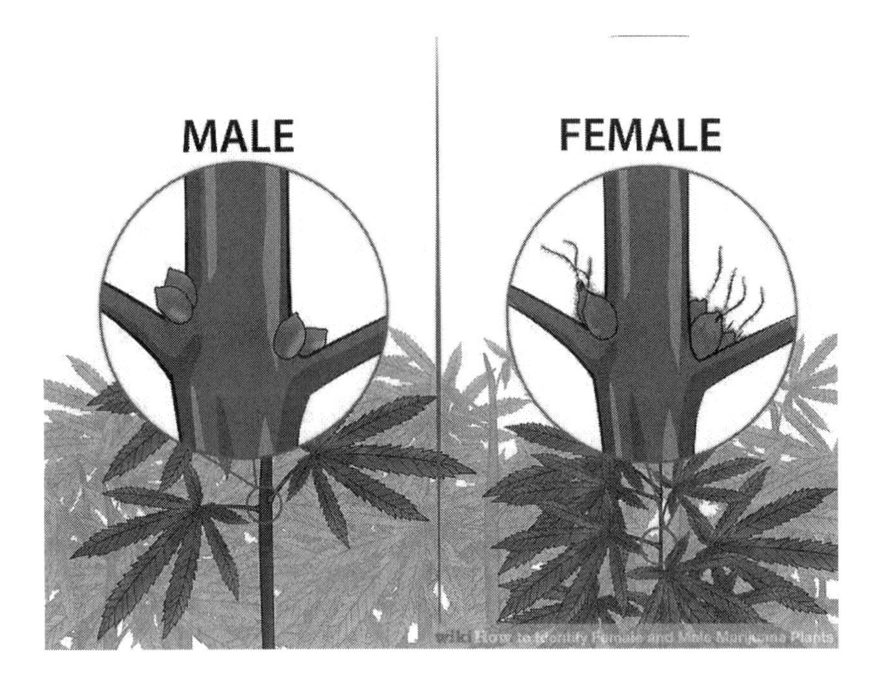

Getting The Light Right

Light is one of the most important elements for any plant to grow. Even humans need light to stay healthy.

But understanding the light cycle is essential for successful growth of your marijuana plants. It doesn't just help them to grow; it can make a substantial difference to how well it grows in different stages of its life.

Look at this way, in nature a plant is told when to start flowering by the changing light of the day. As the hours of light in the day starts to decrease the dark hours will increase and your marijuana plant will know it is time to start flowering.

You generally have little control over this as the plant will respond to the amount of sunlight each day. However, when you are growing marijuana plants inside you have more control over how much of the day is light and how much is dark. You can lengthen or shorten the hours of light your plants get to trigger flowering when you want it.

The purpose of generating flowers is to create pollen and attract insects to transfer this pollen to a new location; allowing the plant to continue to live. This light trigger is essential as it helps seeds to grow.

Here are some methods of lighting you can use:

High Intensity Discharge (HID)

HID grow lights are the most popular grow lights and offer a consistently great yield.

Metal Halide (MH) creates a blue light that's great for encouraging growth in the vegetative stage while

High Pressure Sodium's **(HPS)** yellow light encourages more growth in the flowering stage.

Light Emitting Ceramic (LEC) creates a natural white light that can be used throughout the grow cycle -from seed to harvest and produces UV rays. It takes the best of the previous two bulbs and creates an efficient hybrid.

Sample Setup
250W HID yields ~1-2oz per month
400W HID yields ~1.5-3.5oz per month
600W HID yields ~2.5-5oz per month

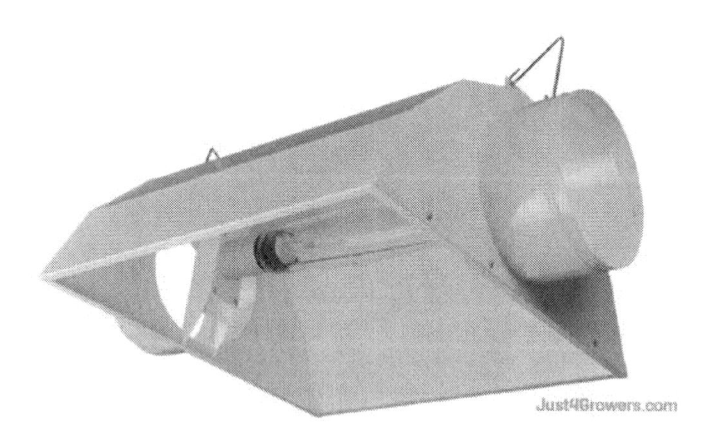

Fluorescents

Fluorescent lights are great for small spaces and to get plants started. They don't burn the plant that's under it. They can be used until the plants are about 24inch tall, and at a distance of 1-4inch. Another big benefit of using fluorescents is the low impact they make on your electricity bill. There are two types:

- T5
- CFL

T5 lights are mostly used for small plants and seedlings and causes them to grow short and wide. In the flowering stage, the T5 lights must be kept close to the buds to produce a decent yield. Look for High Output (HO) bulbs to get the most lumens for your bulb.

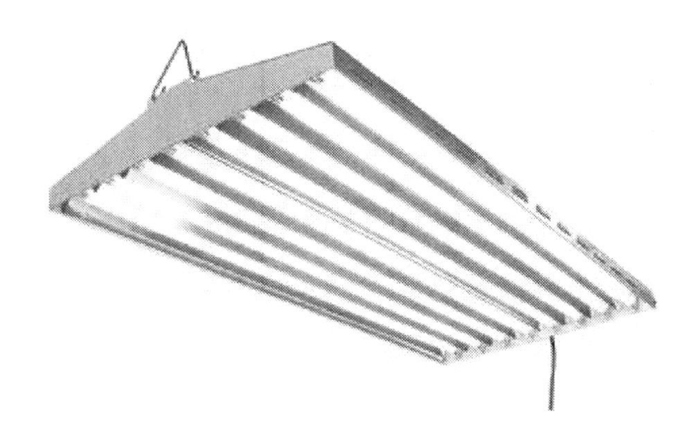

Sample Setup

For vegetative growth use bulbs that are labeled "Cool" or "Cool White" and 6500K.

For flowering growth use bulbs that are "Warm White" or "Soft White" and either 2500K or 3000K.

CFLs can also be kept close to the plants allowing you to cut back on the total amount of space you'll need. These lights can conveniently be used from seed to harvest but won't be able to produce the biggest yields.

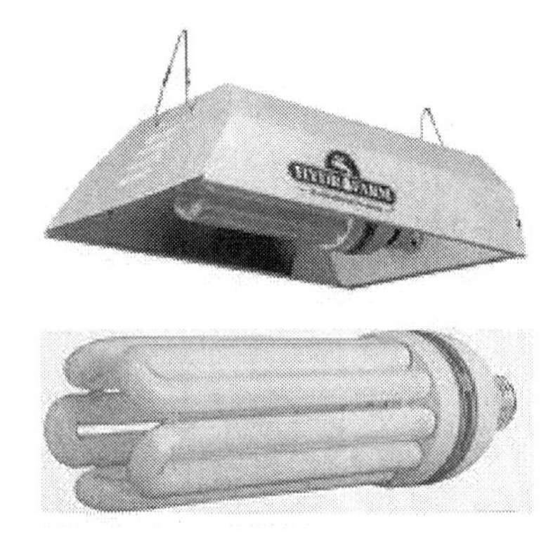

Sample Setup

2 x 40W Daylight CFLs

2 x 42W Soft White CFLs

Equals ~1oz of bud for each 150W of CFL

Light Emitting Diode (LED)

LED lights are the latest in lighting technology and are constantly being developed. They feature excellent yields for the same amount of electricity as CFLs, a customizable spectrum of light, and they don't emit too much heat so you don't need to invest in an expensive exhaust system. In fact, many LEDs come with a cooling option like built in fans to push away the heat from plants to help create and maintain a stable temperature.

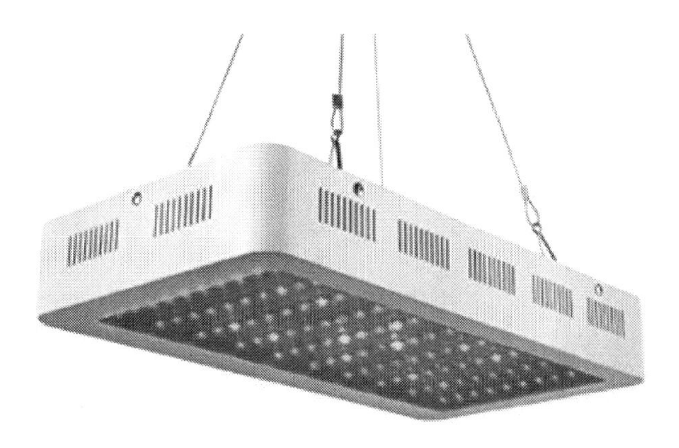

As great as their yields can be, it's been difficult for growers to find the right light spectrum, size of the diode, and angle configuration for each model to grow effectively because they can be so different. It's also important to note that LEDs must be kept at a distance of about 12"-18" from the tops of plants to avoid burning. Until the kinks are worked out and the process is streamlined, there are a lot of unknown risks as much as there are perks to using LEDs.

You can also consider a combination of LEDs and HPS bulbs to improve the overall quality of your buds. LEDs are commonly thought to increase the potency of your strain while HPS bulbs produce better looking buds.

Sample Setup
2 x 125W LEDs

Achieving the Best Growth

In fact, it's not the amount of sunlight that a plant gets which triggers its seed producing response, it's actually the amount of darkness it's subjected to.

The more hours of darkness your plant receives the closer it will think winter is, this will encourage it to start flowering.

In short, keep it green and bushy by keeping the light to a minimum of 18 hours per day. The key level of darkness is 12 hours, once your plants have this they will start to flower. There are slight discrepancies depending on the age of your plant:

- Seedlings need 16 – 24 hours of light and no more than 8 hours of darkness.
- Clones need 18- 24 hours of light and no more than 6 hours of darkness.
- Vegetative plants need 18 hours of light and 6 of dark.
- Flowering need a 50 / 50 split, 12 hours of light and 12 hours of dark.

Photochrome Red And Far Red

This term relates to the color pigments of your leaves and is actually the light receptors in your marijuana plant. They are designed to absorb red light and then react to it, or the lack of it.

Photochrome Far red responds to the red part of the light spectrum. To manipulate this light receptor, you simply need to use lights with far red color in them.
This is important, the moment there is no red in the light the far red will start turning into standard red. Once the red has the advantage over far red the flowers will start to open!
However, while far red converts to red slowly, the process of converting standard red back to far red is exceptionally quick; this is why it is essential to have at least 12 hours of darkness!

How Infrared Works

Infrared operates at the far end of the light spectrum; much of the light emitted is not visible to the human eye. It does not contain far red. This means that infrared can be used to make your plants think it is dark; encouraging flower growth and allowing you to obtain the flowers which you need for your medicinal marijuana.

In fact, the real benefit of infrared is that it can be put on for 6 hours and your plants will think they have had 12 hours of darkness! There is a distinct advantage to this approach; your plants will still be able to get as much as 18 hours of sunlight, allowing them to keep growing while they start flowering.
The result is bigger plants and a better yield.

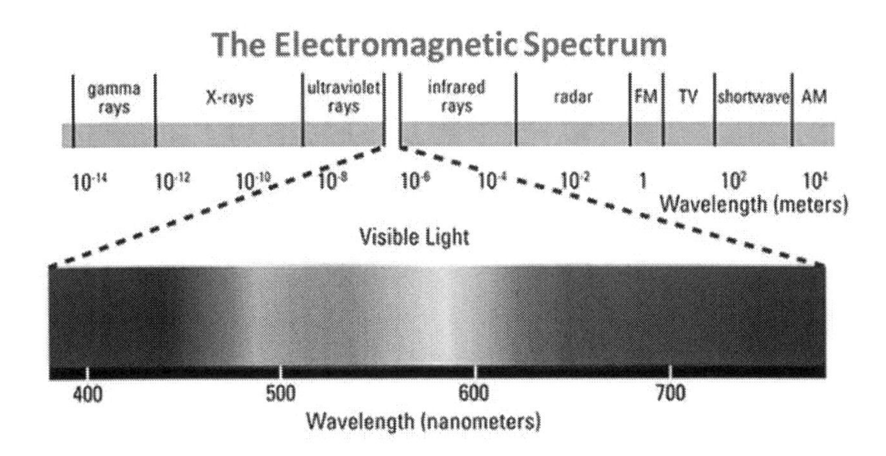

How UV Fits In

Ultraviolet light is another light range that is invisible to the human eye. This is at the other end of the spectrum, between visible light and X-ray light.
There are 3 types of UV light:

1. **UV –A (320 – 400nm)**

This is the least harmful and is often found in nail lamps and other similar products.

2. UV – B (290 – 320nm)

Long term exposure to UV-B can cause damage to cells in plants and humans; the majority of this is absorbed by the ozone layer. This is the same UV that burns your skin.

3. UV – C (100 – 290nm)

This can sterilize most things but is fortunately caught by the atmosphere before it reaches earth.

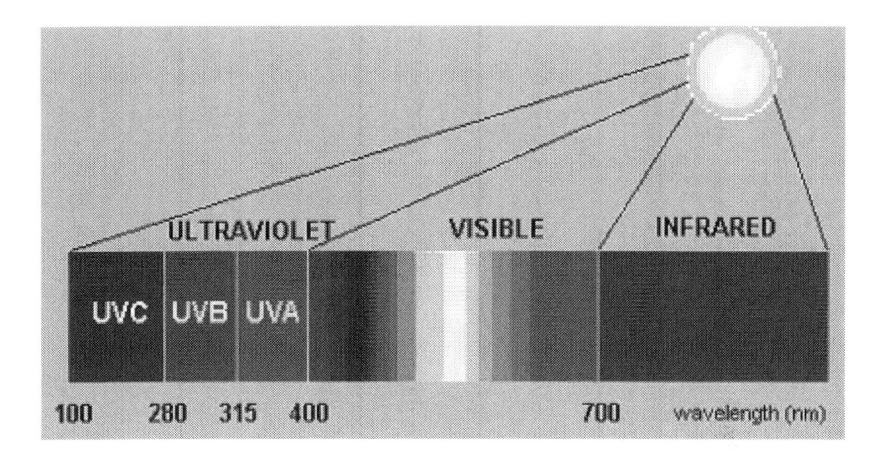

UV-B is important to plants; it triggers a reaction which makes them grow towards the light. It is therefore useful to expose your plants to some UV-B.

But, UV-A is far more beneficial. It will increase the yield of your plant, helping them to grow faster and it will increase the number of nutrients in your plant, ensuring it tastes better. It can also kill microbes on your plants which could be harmful to the growth of your marijuana plants.

In short, a UV light will help your plant to grow faster and larger.

To successfully grow your plants inside you need to use UV lights. They don't just improve the yield; they allow your plants to have enough light. Without UV lighting your plant will not grow properly and will flower too early, making it effectively useless.

Indoor plants will not see enough sunlight to grow; that is why you need UV light on them for at least 18 hours a day. This will ensure they have strong roots and produce a good yield.

Top Tip: If you don't have enough UV light your plants will start to get thin as they search for the light source.

Your lighting system must cover the full spectrum of lights. A grow light is best as this includes UV lighting and the full spectrum of light to keep your plant happy. In the early days of growing you should be focused on building the plant up; you can leave the grow light on all the time to boost their growth.

Kick starting the Flowering Stage

It is possible to kick start the flowering stage and it can improve the yield by shutting all light off for 36 hours. This will dramatically reduce the amount of far red in the plants and start them flowering. However, don't forget that consistency is the key. After the 36 hours you must go to a 12 hours' light and 12 hours dark routine or 6 hours of infrared.

You also need to ensure that your growing room is completely dark when the lights are off. Step inside, shut the door and turn off the lights. After a few minutes you should be able to see nothing. If you can see your hand, there is still some light present. It is best to make sure this is eliminated before you start the flowering stage.

Caution

You must not have your lights too close to your plants; this will cause them to burn. You can also consider a back-up lighting system should the power to your house go off. While it might not have much effect during the flowering stage it is likely to restrict the growth and yield if you are in the vegetative state.

Fortunately, a camping light or two can be enough to persuade the plants not to start flowering, just make sure you have them and batteries to hand!

Getting The Soil Right

Anyone who has tried any type of gardening will understand the importance of getting the soil conditions right. The soil is where your plant will gather the necessary nutrients to grow. There are also different types of soil; texture and drainage can make a huge difference.

A good soil for your marijuana plant will have a light texture and be fairly good at retaining water. If the soil is heavy the plant roots will struggle to spread and there will be too much water for the healthy development of the plant.

In general, it is not advisable to simply go outside and dig some of your soil; it will probably not be beneficial for your marijuana plant. Instead you should look at purchasing a potting soil. This is soil which has been special formulated to provide young plants with all the nutrients they need to get them started in life. The alternative is to get soil compost; this will also be rich in nutrients.

Here are some ingredients you could have in your soil:

- Perlite
- Bat guano
- Bone meal
- Composted humus
- Earthworm castings
- Pumice
- Kelp

Perlite - This is a common addition to soil and one that you could easily add to a standard bag of soil purchased in your local garden center. It looks like little white rocks and it will increase the ability of water to drain through your soil, preventing you from overwatering your plant. In addition, it encourages oxygen into the soil for your plants. However, you shouldn't use too much of this if you are planning to add nutrients to your soil; keep it to 10% of your soil mix.

Vermiculite - This is a good addition if you are losing water too quickly in your plants. A little will help to slow down drainage and can make the soil heavier. But it will also restrict the addition of oxygen to the soil.

Worm Castings - These are full of nutrition that is essential and extremely beneficial to your marijuana plants. In fact, there are no real downsides to adding this to your soil. It can make a valuable addition on a regular basis to encourage good microbes and provide nutrients for your growing plants.

Bat Guano - This is a great way to add nutrients into your soil. Simply sprinkle it on the top of the soil and keep it moist. This will encourage bacteria to feed on the guano and break it down to create the nutrients your plants need. It can be used throughout the entire growing cycle.

Bone Meal - Bone meal is high in phosphorous and calcium which are great for your plants as they start to flower. However, it a slow releasing fertilizer so it's best to be mixed in your soil when you start potting. The nutrients will then be available when they are ready to flower.

Composted Humus - This is basically any type of compost which has naturally decomposed. Because it is natural materials which have been broken down it is full of nutrients which will benefit your plants.
It also benefits your plant by helping the soil to retain moisture and oxygen. The result is better quality soil for your plants.

Pumice - Pumice is a type of volcanic rock. Because it is porous it is good at holding water and allowing air flow. It works in a similar way to perlite and should be mixed with your soil.

Kelp - Kelp is a plant in itself and as such contains all the nutrients that every plant needs to survive. This makes it a great addition to your soil for encouraging plant growth.
But, it is also extremely beneficial as it is excellent at repelling slugs and other pests from your marijuana while keeping the soil moist. You can cover your soil throughout the growth of your plant but only after it has germinated, you don't want to suffocate it as well as the weeds!

Once you have got your soil sorted you now need to consider fertilizers. If you are planning on repotting your plants regularly then this might not be necessary as the fresh soil will have its own nutrients. But, if this is not the case fertilizers are essential to ensure your plants have everything they need to grow big and strong!

It is possible to mix your own fertilizers. However, unless you are certain about what is already in your soil this can be a dangerous process. Homemade fertilizers can react with chemicals in the soil and cause a detrimental effect to your plant.

As a beginner you should purchase one of the many fertilizers on the market. This will help you to choose the right one for the stage of plant growth. A general fertilizer works well for seedlings but flowering plants must have balanced nutrition. Once you know your soil types and understand the needs of your plants you can use the following as your own fertilizer:

Chicken Manure - A little chicken manure can go a long way when looking after you plants; it's full of nutrients that are essential for all stages of plant growth.

Vinegar - Did you know that one drop of white vinegar on baking soda will release carbon dioxide; which your plants thrive on.

A good way of doing this is to put the vinegar in a plastic bag and hang it over the bowl. Prick a tiny hole to allow it to drip out slowly. But, you must do this in a way that keeps the carbon dioxide in the space; this is not good for outside growing. There will also be a heavy smell of vinegar which you won't want others noticing.

Kitchen Waste - All the organic food waste which you usually chuck away can be put into a pile to encourage them to decompose. Once they start to decompose you can put them on the top layer of your soil; releasing all the nutrients your plant needs when you are watering them.

Getting The Pot Size Right

As with any plant the bigger the pot the larger the plant will be able to grow. This means you need to know which type and strain of marijuana you are growing to ensure you have the correct pot size.

In general, these sizes will apply:

- 2 - 3 gallon container will yield a 12″ plant
- 3 – 5 gallon container will yield a 24″ plant
- 6 – 8 gallon container will yield a 36″ plant

A good size for most cannabis plants is 3 gallon pots, this will allow them to grow and establish the deep roots they need to have without taking up too much space. After all, you'll want to grow as many plants as possible (within state laws, of course).

You can go larger if you have the space. It will not guarantee bigger plants but it will offer them the potential. The right light, soil type and nutrients will also be necessary. Please note it is best to have one pot per plant, this will help prevent any issue with one plant affecting another and ensure they all have the opportunity to grow properly.

You should also ensure that there are drainage holes in your pots; this will prevent water from collecting in your pot and drowning/rotting your roots.

You can also start small and repot as necessary. However, re-potting does place stress on your plants and can be time consuming. You'll also need to ensure you do it at the right time. As soon as the roots start to poke through the bottom of the pot it needs replanting, or it will become pot bound and grow no bigger.

Other Pots to consider

The standard flower pot can work well but you'll find there are a couple of other types of pots which can also be beneficial:

Smart Pots - A smart pot is made from fabric and encourages air to be in direct contact with the roots as soon as they touch the sides. This will encourage them to grow more roots and the plant to grow larger and stronger. But, the plant will dry out quicker, you'll need to buy larger pots to account for this.

Air Pots - These plastic pots have openings on the side to offer the same effect as the smart pot. The only real difference is that they are sturdier.

The Hydroponic Method

You might not have considered the idea of growing your marijuana plant with hydroponics but it's a common and successful approach.

However, you should be experienced at growing marijuana plants before you attempt this method. It requires a solid knowledge of the nutrients your plants need at each stage of their growth. In fact, there are several different options when using hydroponics, the most advanced will of course, increase your yield.

Hydroponics simply means growing in water, there is no soil needed. Many plants do this in nature. You'll need a pot with enough substrate for the plant to stand up straight. The roots will be free and easily able to absorb the nutrients you put in the water.

Growing indoors a hydroponic system will give a higher yield and the plants should grow faster. But, it is more expensive to set up in the first place. Each plant is placed in a pot with substrate; these are good substrate ideas:

- Rockwool
- Clay pellets
- Coconut Fiber
- Perlite
- Vermiculite

All hydroponic systems work on the principle that soil is not needed, just a substrate and nutrient filled water. Here are 2 good examples of hydroponic systems:

Passive or Closed Circuit System - From each pot you'll have a wick which extends below the pot and into a pool of water. This will need to be regularly topped up with water and nutrients. The plant will take all the nutrients directly from the water. This is a good way of starting hydroponics. But, the water is static which increases the chance of a bacterial infection and, if this happens, it could easily damage many or even all of your plants.

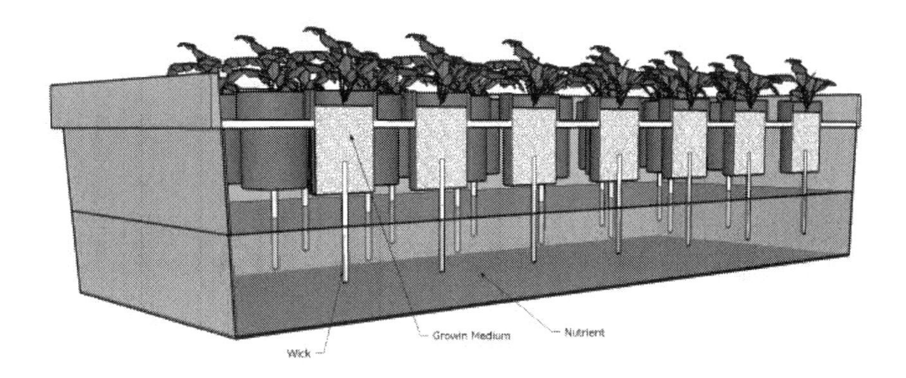

Ebb & Flow - You'll need two containers. One will act as a reservoir for the nutrient and water mixed. This is pumped in the bottom of another container which has an overflow pipe to stop the water getting too high. The pup comes on periodically on a timer to ensure the nutrients are circulated efficiently.

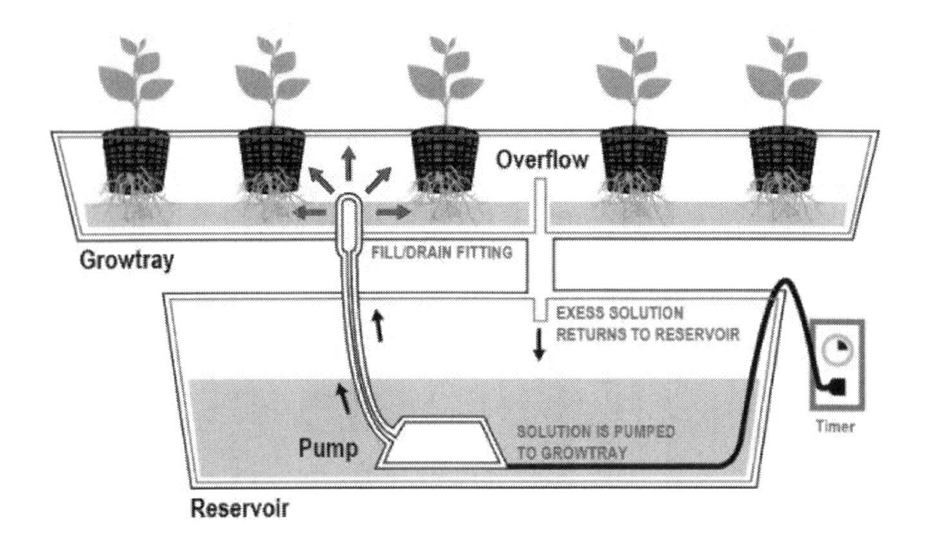

The plants sit just above the water level with the roots dropping into the water; giving them all the hydration and nutrients they need.

Aeroponics - This can be considered the most advanced form of hydroponics; it is also the most difficult to get right.
The roots of your plants will simply be hanging in the air (inside a pot that holds the plant in place). You'll need a large tank to act as the reservoir. This can be one third filled with water and the right nutrients for your plants. A submersible pump is placed in the water and connected to a spraying system.

Every spray should be located near the roots of the plants. You then set a timer to ensure the pump sprays the plant roots periodically. Without the spray the roots will quickly dry and the plant will die; this is why it is the most difficult system to get right and one you should aim to work up to.

On the plus side, this will give the biggest yield in the shortest space of time! The takeaway from this is that whether the roots are in light soil with good drainage or in the air, misted by a nutrient spray, you need to make sure every plant is receiving the nutrients it needs.

Beginners should start with pre-packaged nutrient mixes, high in nitrogen for young plants and full of potassium, oxygen, phosphorus and nitrogen for all plants. Keeping the nutrients, soil and lighting right will be a huge step towards a successful harvest.

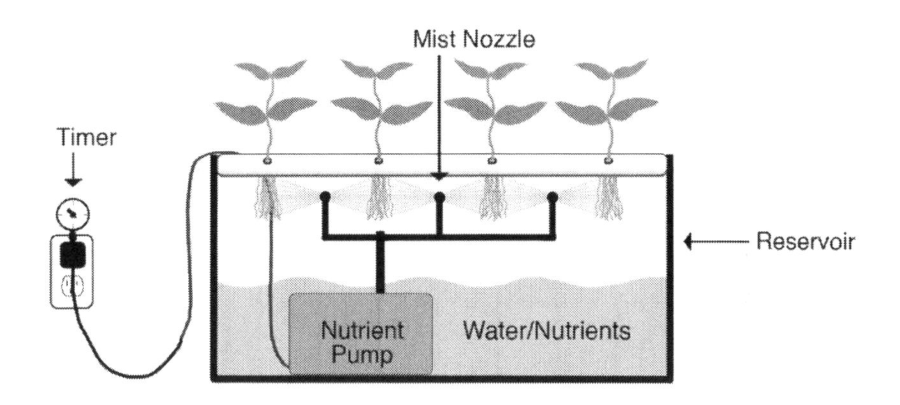

Although hydroponics is an interesting way of growing your plants, one could write a whole book about this topic. That's why I decided to not go into much detail about it in this book.

Atmospheric Conditions

At this stage, things should be starting to come together. You've chosen the type of marijuana you wish to grow, your pot size and the type of system you wish to use to grow your marijuana. For clarity, if you are a beginner then it is best to start with a pot, good quality soil and some fertilizer. Providing you've chosen the right light set up and understand the light cycles you're nearly ready to start planting.

But first, you need to understand the importance of getting the humidity right, how to ventilate your crop and, perhaps the most concerning of all, how to prevent the aroma of marijuana penetrating the air. Even if you have just 6 plants this aroma might be enough to attract others that would like your crop for their own reasons!

Setting The Humidity Level

You are probably already aware that humidity simply refers to the amount of moisture there is present in the air. The more there is the more humid the air.

The marijuana plant is not keen on high levels of moisture; this is good as modern heating systems tend to dry the air in your home. However, nature is all about balance. Low humidity levels equate to high evaporation pressure. This is good as it helps your plants to absorb vital nutrients. However, if the humidity becomes too low then the plant will assume there is a problem with water and start protecting itself from dehydration.

Unfortunately, this means it will no longer absorb water and will not be able to grow!

When you have seedlings they need a humidity of 70%.
As each week passes you can reduce this by 5%. This is because the roots are growing and increasing the plant's ability to collect water.

After 6 weeks you should be at 40% which is where you should keep the humidity level for the rest of their growth. Humidity is partly controlled by the temperature of your grow room. It is therefore essential to have a humidity meter and consistently monitor the level; this will ensure you can react appropriately if the humidity levels change.

How to adjust humidity
You need to know the current humidity rating. It's best to use a hygrometer for this (analog or digital).

It is worth noting that the humidity is affected by the temperature outside the house. For instance, temperatures outside below 15° Fahrenheit your humidity reading will probably be about 35%. It should increase by 5% every time the temperature rises by 10°F.

Air-conditioning or a dehumidifier. These are a great way of removing moisture from a room, allowing you to bring it down to the most appropriate level for your needs.

Cat litter can be your best friend if the humidity gets too high. Simply spread some in a tray and leave it in the growing room. It attracts moisture and will lower the moisture level in the room.

Add moisture by having water in the room in open bowls, it will evaporate into the room; increasing the humidity levels.
Another option is to use a humidifier which will push moisture into your room, helping to boost the moisture level.
You can also use a vaporizer; this uses warm water and vaporizes it into the air, boosting the moisture content, if needed.

Add ventilation, this can reduce the humidity level if the outside air already has a lower humidity rating. A fan, coupled with an atmospheric controller can make a huge difference to the ease in which you can control the humidity levels.

De-leaf some of the plants with the most leaves on. Excessive leaves can increase the humidity in your growing room; this is particularly true when you have limited airflow.
These are all great methods to boost or decrease your moisture levels, however the key to this approach is making sure that you monitor the levels regularly and adjust them slowly.

Poor Humidity Side Effects

Unsurprisingly getting the humidity wrong will affect the growth rate of your plants. But this is not the only sign that there is an issue with humidity:

White powder - This is a fungal disease that only arrives when the atmosphere is too humid. A good airflow system can help to prevent this.

Bud Rot - If the insides of your buds are white or brown with mold then you have bud rot and your crop is effectively useless. This is incredibly frustrating and not something that will be an issue if you monitor the humidity rating properly.

Nutrition - If your plants start to look like they have yellow or burnt tips then they are effectively consuming more water than they should be; this is usually because of low humidity.

Increase the humidity levels immediately.
Don't forget the right humidity levels will encourage maximum growth!

Plant Stage	Humidity level
Seedlings	70 – 80%
Vegetative	40 – 60%
Flowering	40 – 50%
Final flowering	Less than 40%

Choosing The Right Ventilation System

Ventilation is often a difficult issue. The basic truth is that no matter how many or how few the number of plants you are growing; you are likely to be growing them in confined conditions.

There is good reason for this; you don't want everyone knowing what you are doing. Even when growing them for medical reasons many people do not realize this is an option and are likely to report you. This will give you unnecessary hassle and highlight to others that you have marijuana.

You need ventilation for several reasons:

Toxin removal - Marijuana is no different to most plants in that it pushes toxins out of itself through the leaves. Air flow helps to remove these toxins from the plants. Without this they can stay there, preventing the plant from pushing more out and even encouraging the growth of mold.

Humidity- We already looked at humidity and understood the importance of getting this right. However, as we mentioned, air flow can help with distributing the same humidity level across the room.

A fan can simply move air round your room, helping to decrease the chance of mold forming on your plants. It can also be used to bring air inside from outside the building; this can lower the temperature to reduce the humidity levels. This approach is especially effective if you add a temperature and humidity control device as the fan can switch on and off automatically.

Creating the Ventilation

The most comprehensive method of creating ventilation is through the use of several items:

- An intake fan, this sucks air in from the outside.
- An extractor fan; this pushes the air out of your grow room and into the outside world.
- An interior fan, to move the air across the plants. This might not be necessary if you only have a few plants.
- Some sort of air filter device.

The intake fan does more than just bring air in, it pushes it in at the same rate the extractor is removing it; this allows the air pressure to remain the same preventing any disruption to the growing cycle of the plants.

Dealing with the aroma

Of course, if you've ever had any experience with growing marijuana then you'll know it has a distinctive smell which is relatively easy for others to detect. While ventilation is essential, this will push the smell of the plant into your house; not something you are likely to want to risk!

First, you need a fan to remove the air. You should already have this as part of your air flow and humidity control measures. Moving the air outside will reduce the potency of the aroma from your marijuana plants.

But this is not enough; the aroma could still give you away and attract unwanted attention. There are several ways of dealing with the aroma:

Carbon Filter - This is the perfect addition to your extractor fan. All the air leaving your room should be pulled through the fan. In the process it will need to pass through the carbon filter you have fitted.

Choose a filter which fits perfectly into your exhaust ventilation system (there are sets). The carbon will attract the aroma and hold it, preventing it from getting into the outside world. You do need to have a fan and you need a quality carbon filter to do the job properly. Searching on eBay or Amazon for "carbon filter fan" will do the job.

Negative Ions - A more advanced method is to use a negative ion generator. This charges the particles in the air and effectively gives these particles the ability to neutralize the odor in other particles; eliminating the aroma of marijuana in the process.

A carbon filter is more effective but this is a great alternative.

HEPA - This is an alternative type of filter which can be used in your grow room. However, it still requires air flow through it. But, the fan does not need to ventilate outside the grow room; making it an attractive option in some scenarios.

Choose the ventilation system and aroma protection system that suits your needs and budget this will ensure you monitor the humidity and your plants will start growing well. It is worth noting that a Hepa filter cannot filter very small particles. You will need a carbon filter for this.

Controlling Temperature

Like all plants it is important to get the temperature right. In fact, temperature is the key to controlling humidity and ensuring you have healthy plants. In basic terms you can simply monitor your growing room by how you feel; if it's too cold for you then it's too cold for you plants. Equally if you think it's hot so will your plants.

You'll then need to take steps to control the temperature. Of course, adding and removing heat will affect the humidity in the room and this must be accounted for. This is why most advanced marijuana growers have humidity and temperature monitors. Controlling these will help to ensure your plants grow large and healthy.

The best temperature for growing marijuana is between 17 - 30° Celsius (62 – 86° Fahrenheit). To be specific:

- Vegetative State – 21 – 30° C (70 – 86°F)
- Flowering State – 17 – 27° C (62 – 80°F)

It is worth noting that the temperature should be lower when they are in their dark stage than the light stage, but still within the range indicated above. This helps to stimulate the natural conditions they have in the wild and encourages the best growth.

Removing Hot Air

Hot air rises so you need to remove some of the air at the top of the room to allow the cooler air to circulate. This can be incorporated in your ventilation system, simply having the extractor pipe at the top of the room or the window.
There are a few extra measures you can take to keep the temperature from getting too high:

Bulbs - Light bulbs give off a huge amount of heat; they can easily raise the temperature in a sealed room and if they are too close to your plants can actually burn their leaves. You can remove excess heat by having ducting running across the ceiling of the room. There will need to be open points next to the light bulbs, allowing the hot air to rise into the ducting. A small fan will then draw the hot air through and expel it where you want.

Expelled hot air can be directed towards the rest of the building or sent out of the growing room through a window to the outside world. However, you'll need to make sure there is no chance of cold air being blown in the pipes and no light is let in where the duct leaves the growing room.

Boosting The Temperature

It is worth noting that if it is too hot your plants will grow much more slowly but they will not usually die or suffer any other issues. However, if they are too cold they can go into shock and even die; this makes it far more important to keep the temperatures up.

In general, your lights will help to ensure the temperature is at least satisfactory if not too hot for your plants. However, what happens when you are in the dark cycle and you wish to keep your temperature within the right range?

Insulation - You can insulate every part of your growing room. This also prevents unnecessary and unwanted light from getting in. More importantly it will make it much easier for you to control the temperature of the room.

Fans - The heat generated by the lamps should be spread across the room; this will ensure the whole room is an even temperature. This can be easily achieved through the use of fans. If you need to bring in cold air from outside as part of the ventilation system. You should mix these with the hot air at the top of the room, this will help to keep the temperature balanced within your room.

Space Heaters - Using an electric space heater with built in thermostat can do wonders. Be careful not to point the heater directly at your plants. Space heaters will warm your grow room up quickly, upping the humidity. Keep a close eye on these two variables.

Another more advanced option is to run several pipes of hot water through the growing room, ideally above the plants. These can be connected to a standard central heating boiler and will keep the temperature comfortable. In effect it is a central heating system without the radiator.

Reverse Growth

It is also possible to reverse the growth routine. Have the light time during the night period; this will help to maintain the temperature even on a cold night. Your dark time can be during the day which will help the temperature to stay high, just don't enter the room and let any light in when you are attempting to get your plants.

To give you an idea of the heat produced by a HPS lamp, a 600-watt lamp needs a minimum of 1m x 1m x2m / 3ft 4" x 3ft 4" x 6ft 8"; or the temperature will quickly go above 38°C. (100° F)

This can also help you to calculate the size of your extractor fan. This can be done by calculating the cubic feet per minute of your growing area.

To do this you need to calculate the cubic area of your growing room; this is the length multiplied by the width and the height. You then multiply this figure by 2 if you have short straight ducting. If you have longer ducting and bends you need to multiply the cubic area by 3. The resulting number will give you the fan size that you need.

Watering Systems

Water is the most important ingredient for all life. Humans, plants and all other living beings are composed of at least 50% water.

This means that an effective watering system for your plants is essential. But, there is such a thing as over watering your plants. To ensure you get the quantity right you need to understand the watering cycles, learn how to detect too much or too little water and understand the irrigation systems available to you.

Watering Cycles

The watering needs of your plants will depend on the pot size. Larger pots hold more water and will therefore need to be watered less often. Seedlings should generally be watered 2 or 3 times a week; assuming they are in small pots. Vegetative and flowering plants can be watered just once a week.

Detecting Too Much or Too Little Water

However, these are only guidelines. The size of your pot, humidity, temperature and even the soil you have chosen will all affect the needs of your plants. Assuming you have followed this guide and used a light soil, got the temperature and humidity right and established a good air flow and light schedule then once a week should be enough. But, you need to check your plants regularly and look out for these signs:

Soil Dryness - This, of course is a good sign that the plant needs to be watered.

Drooping Leaves - If they are hanging down and start to pick up after you have watered the plant then they are being under watered, you need to do it more often. If you don't the leaves will start to yellow, showing that there is a nutrient deficiency. The stems will also be drooping down. If you know the weight of your pot with dry soil, you can lift it to see if there is still water left in the pot by comparing its weight.

If you water them, the leaves will be back to normal in a few hours.

Bloated Leaves – If the leaves appear bloated and thick but droopy then it is likely that the plant has been overwatered. This might seem the same as under watering (because of the droopy leaves). The main difference between the two is that the stems of an under watered plant hangs down and is soft while the stems of an overwatered plant still points up and is strong.

Tips Curling – When the tips of your plants leaves are starting to curl under themselves then you know it has been under watered for some time. This is a sign of nutrient damage and needs dealing with quickly.

A plant recovers quicker from under watering than from overwatering.

Getting The Watering Right

The simplest way of watering and monitoring your plants is to do it by hand. This will allow you to feel the soil and see how quickly the plant responds. The soil should be dry to the touch but the plant should not yet be wilting. You can then pour water on until it pools on the top of the soil. Stop immediately and the water should disappear; draining through the soil and into the collecting dish.

If you are adding nutrients to your water it is best to add a little extra water at this point, this will ensure there are enough nutrients in the soil and encourage the roots to move downwards. If your pots do not appear to be draining and the water is staying on top of the soil then you have a problem! You are effectively over watering your plants. Check the soil has good drainage, add additional perlite if this helps.

You should also make sure the pot has holes in the bottom to allow the water to drain. If necessary, you can repot the plant but it will need to go into a bigger pot to encourage it to continue growing.

A clever tip to help you know when they need watering is to fill a pot with the same growing compound as your marijuana plants. If the marijuana pot is heavier than it has enough water, if its lighter it needs water (also compensate for the weight of the plant itself).

Irrigation Systems

The more advanced you get with your marijuana growing techniques the more interested you will become in ideas to create irrigation systems which take the strain off you. Watering all your plants can be a time consuming project.

The Drip System

This is the most basic form of irrigation systems. Simply fill a bag with water and hang it over your plant. Prick a very small hole on the bag and allow it to drip onto your plant. While this method is hands-free, you don't actually control the flow.

A more advanced version of this involves running a small pipe above your marijuana plants. The pipe can be rubber, plastic or metal but it will need lots of tiny holes in. The pipes are all connected together and then connected to the mains supply (a bucket with a pump and dissolved nutrients). A simple twist of the tap will send water through the pipes and out the small holes.

Go Electronic

A step up from this is to attach the pipes to your water supply via an electronic valve. This can be connected to a timer and allows you to set when the plants should be watered. You won't need to do anything to ensure your plats are watered, just verify that they are not over or under watered!

Tracking and Adjusting the pH

pH is a measurement of the amount of free hydrogen and hydroxyl ions in water. If there's more hydrogen ions, the water is more acidic. If it has more hydroxyl ions, it's more basic.

How do you do test the pH? You can figure out the pH of your water you by picking up a basic pH tester kit and adding different solutions, called a pH Up and pH Down, to nudge the pH to where you want it.

Hard vs Soft Water

The amount of pH Up or Down you add to your water will vary depending on the water you start with. Water quality is generally considered "hard" or "soft". Soft means that there's not anything in the water. It's been treated to the point that the only ion present is sodium. Hard water contains more minerals like chalk, calcium, lime, and magnesium. If you're using soft water, you only need a small amount of fluid to adjust the water's pH because there are no other minerals to buffer the pH. If you're using hard water, you'll need to add more pH to account for the additional materials.

pH Up

pH up is much weaker than pH Down so you will need more of it to alter the overall pH. As a guideline, 2-4mL per gallon of water will raise the pH by 1 point.

pH Down

1mL per gallon of water will generally reduce your overall pH by 1 point.

Your initial adjustments will involve a lot of "guessing and checking". You'll add a little and then check the pH and adjust from there. It's helpful to make notes on the total pH Up and Down you've added to speed up the process later. You want to keep a root pH in the range of 6.0 to 7.0 if you're planting in soil. If you're planting it in hydroponics between 5.5 - 6.5 is the goal. There's not one number you need to keep it at. A range is actually good as different nutrients are absorbed better at different pH's. Tap water is generally around 7.5 Ph.

More Tips for pH Management

Gently shaking the water evens out the ratio of nutrients and the pH Up or Down. The extra oxygen is also good for the plant's roots.

Use tap water or mineral water to create that extra buffer in the water. This helps to keep the pH relatively stable and easier to manipulate.

From Growing to Harvesting

The turnaround time for growing your own cannabis is generally around 4 months. If you're working with clones, it can take somewhere between 2 and 2 1/2 months. We'll break down the stages from growing to harvesting next.

Germination

Germination is the process of getting your seeds to sprout, which can take between 12 hours and a few days. The seeds need three things to properly germinate: warmth, darkness, and moisture. When done properly, you'll see a little white tendril pop out of the seed, that's the tap root. Next we'll go over three ways to get your seeds to germinate: a seeding plug or starter cube, planting directing into the soil, or soaking overnight.

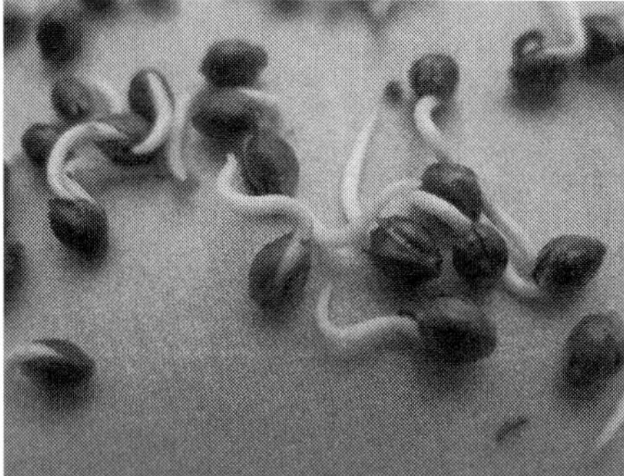

Before you begin, make sure your hands and workspace are clean. Especially if you're a smoker, nicotine can damage cannabis seeds and plants so you'll want to be clean of that.

Using a Seedling Plug or Starter Cube

To start them in a seeding plug or starter cube, simply place the seed in the plug or cube and water as directed. There's even a hole for you to insert the seed and get things going.

Planting in Soil or Soilless Mediums

For soil or soilless mediums, you can also plant directly into the final medium. Just plant the seeds knuckle deep in moist soil. The benefit of planting right into your final medium is that you don't have to worry about transplanting and any issues that could arise from stressing the plant.

Soaking Overnight

Finally, you can soak overnight. Using a glass cup and slightly warm filtered water, place your seeds into the water and let them sit overnight. The most viable seeds will float to the top and then gradually sink down. After 24 hours, check in on the seeds to see if the tap root has broken through. Leaving the seeds in water for too long can cause them to drown, so if they haven't germinated after 24 hours place the seeds in a warm, moist spot to finish germinating.

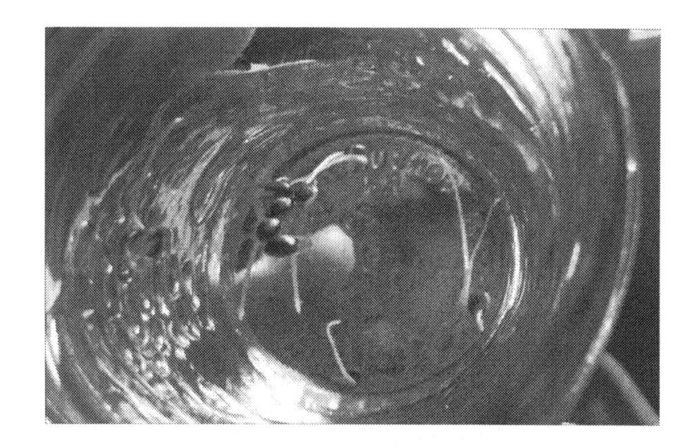

Once the seeds have sprouted, plant them as soon as you can into a bigger container. Don't touch the tap root with your fingers as it can be damaged easily. Plant the seeds so that the root faces downward and is about knuckle deep in your growing medium. Whichever method of germination you choose; it does make sense to keep the new seedling in a small container at first to allow the roots to have access to the most oxygen.

It's in no way a death sentence to plant your seedlings in a bigger container, but they might grow slower in the beginning. Don't forget to poke holes in the bottom of your container to allow for the water to drain out

The seed has officially germinated and begun the seedling stage when the root has fixed itself into the soil and pushes two leaves through to the surface.

Seedling Stage

The seedling stage features round leaves, called cotyledons. After these initial leaves, all others will start to have the characteristics of cannabis leaves. Specifically, the blades or "fingers" of the plant leaves which will increase as the leaves keep growing.

Sometimes the stems are very weak during this stage and can be supported by tying them with thread to a thin wooden stake. This stage can last for 1-3 weeks indoors and 4-6 weeks outdoors. At the end of the seedling stage there will be 4-8 new leaves.

At this time, the seedling requires moderate humidity levels, medium to high light intensity (18 hours), and adequate soil moisture. Lights in the blue spectrum that produce little heat like T5 fluorescents, are a great choice for this stage because they won't dry out the plant.

You can also use HID lights though they will produce more heat and you will need to use the hand test to monitor its temperature. It's also a good idea to have a weak fan blowing on the seedlings to help them get fresh air and to have something to resist to encourage strong root growth.

If your seedling's leaves are hanging down, they are definitely thirsty and need to be watered immediately. Make sure the containers they are in are draining properly as well. This stage of the growing cycle is the foundation for the rest of the plant's life.

The seedling stage has ended when the leaves have the maximum amount of fingers generally 5 to 9 (the number varies from plant to plant), has a stem width of 4-6mm, and a height of 3-4 nodes. You will also notice that the roots will begin growing out of the holes in the container it's in which is another indicator that it's time to carefully begin the transplant process.

Vegetative Stage

The vegetative stage is marked by leaf and stem growth, not bud growth. During this period of 1-2 months, your plant will get bigger and leafier so they can support the weight of the buds. Amazingly, a healthy cannabis plant can grow up to two inches in a day! Keep temperatures in the 70-85° range with a humidity of 40-70%. In terms of nutrients, make sure your levels of nitrogen are high, phosphorus is at a medium level, and potassium is high. A standard vegetative nutrient formula will follow this.

Your plants should be getting between 18 and 24 hours of light a day from a bulb on the blue spectrum during this period. Cannabis will stay in a vegetative state until the plant begins to experience nights that are shorter than 12 hours.
It's important to keep this consistent. Any inconsistencies can stress out the plant and potentially turn it into a hermaphrodite, which will seriously alter the quality and quantity of the buds. Indoor growers can control how long the vegetative stage is whereas outdoor growers are subject to nature.

If you're growing indoors, it's best to force your plants to flower when the tips of the leaves are touching each other. They will continue to grow in the first 2-3 weeks of the flowering stage anyway but you don't want your plants to outgrow your room. In terms of nutrients, make sure your plant is getting plenty of nitrogen.

Cloning

If you want to clone a plant yourself, the vegetative stage is when to do it. Using a clean razor blade, cut a branch off of the stem during the vegetative stage. Cutting a branch during the flowering stage will cause stress to the plant. You want to cut down along the natural angle where the branch meets the stem, this should be at about 45°. Then cut down the middle of this branch, just a little before the end of it, before planting. Cutting the branch twice like this increases surface area and encourages root growth. When transplanting it to a smaller pot or tray and grow medium, you can also use rooting hormones or cloning gels to encourage root growth.

Clones take about three weeks to develop healthy roots. Once the roots are established, it's time to transplant them to their next container. Here they must be under light for 18 hours a day until reaching a height and width that just touches the other plants. Once it reaches this stage, it's ready to move on to flowering just like a plant you had from a seed.

Pre- Flowering Stage

If you're growing outdoors, you can expect the flowering stage to begin as the days get shorter and we head into fall. If you're growing indoors, this is the time to change your light schedule to 12 hours on, 12 hours off so the plant begins to flower. Avoid any light interfering with the actual or artificial "night". That means no street lights or spotlights coming off of a back porch or something similar. If the plant's nights are encroached by light they might revert back to the vegetative state and delay flowering.

During the first few days to two weeks after switching to a 12 on, 12 off light schedule, the plant will rapidly grow in height and branches as the plant grows. This period is commonly called "the stretch". Keep your plants spread out and gently bend any stems down and away from the center of the plant that look like they are "reaching" or stretching very high.
By creating a flat canopy of leaves you will increase your bud yield by as much as 40% because it evens out the light distribution. Keep in mind that you won't be able to make adjustments like this later when the stems harden and become harder.

The sex of your plant will also be revealed as it's getting ready to share its genes. Female plants will begin to grow white pistil hairs. Male plants will grow pollen sacs. If you are growing both, make sure to remove the male plants to avoid pollination.

If you buy seeds make sure they are female. If you buy female seeds, there would still be a chance of having male seeds. So make sure you check if they are all females.

Flowering Stage

You can expect the duration of the flowering stage to last between 6-10 weeks. Drop the temperature down between 65-80° and the humidity to 40-50%. As far as nutrients go, keep nitrogen levels low, phosphorus in a medium to high range, and potassium high. A standard bloom nutrient formula will follow this recommendation. We'll explore what happens in this time period next.

Weeks 3-4 Beginning to Bud

The rapid growth upwards you've been seeing will begin to slow down after 3 or 4 weeks and buds will start to emerge. At this stage, your plants may also start to smell. During this time, you'll need to pay more attention to your plant as its environmental and nutrient needs will change. Do not change your nutrient solution until the plant shows very clear signs of flowering. Check out the troubleshooting section later on if you notice negative changes like yellowing.

Weeks 4-6 Buds Fatten

Now you can expect the buds to become more substantial and all of the white pistils will be sticking up straight. Your plant will no longer be making new leaves or stems so make sure you are watching them carefully for any problems during this time. You may also prune away all the buds that are in shadows or leaves that are dead or otherwise compromised.

Weeks 6-8 Buds Ripen and Pistils Darken

With all of the plant's attention turned from general growth to developing buds, it's more susceptible to pH and nutrient problems. Watch out for yellowing leaves which can be a sign of light burn or a nutrient deficiency. In the final weeks of flowering you will also want to drop the humidity to less than 40%.

On female plants you can expect to see a lot of sticky resin secreting from the trichomes on the outside of the leaves and buds. This resin is THC, the ingredient that medicinal cannabis users seek for its therapeutic qualities. The potency of your plant is determined by the amount of time it's spent flowering and if it's been fertilized, which is why it's important to remove male plants.

After 8 Weeks

By now the buds are fat, the trichomes and the pistils are mature, and your plant should have a very strong smell. Some of the leaves might yellow during this time but as long as the yellowing isn't impacting the buds, you're doing fine. Don't raise the nutrient level to try to combat minor yellowing.

It's a common practice among growers to stop providing nutrients to their cannabis for the last few days or up to two weeks of flowering before harvesting in order to let the plant flush out nutrient build up. If there's a nutrient build up it can create a chemical taste in the final product. Don't flush the plant of its nutrients until the pistils are nearly all darkened and curled.

Harvesting

When to harvest is highly dependent on the strain you've chosen and your personal preference. You can start as early as after two months of flowering or wait longer than four months after initial flowering. The timing of the harvest determines the taste, smell, weight, and effects of your cannabis and should be considered carefully.

The best way to decide when to harvest is to look at the pistils. These little hairs are white when they first appear, but over time they curl and change color. They can turn yellow, pink, red, purple, and brown. A guideline regardless of strain, is to harvest when 70 to 90% of the pistils have changed color, but here's the total breakdown:

Look at the Total Percentage of Color Change

- If 0-50% of the pistils are brown, they're not ready yet.
- When 50-70% of the pistils are brown they can be harvested but are considered young. At this age they will give you a light taste and a mellow high.
- When 70-90% of the pistils are brown they're ready to harvest. This is the peak of the plant's taste and effect.
- After 90% you're edging into it being too late to harvest. The taste is heavy and the effects are compromised. Don't wait any longer to harvest.

Other Tips

In these last two weeks of growing, some growers feed their plants a teaspoon of blackstrap molasses per gallon of water. The sugar helps the buds improve in size and flavor.

Trim the leaves from the buds using small scissors. The big leaves have no active ingredients that would be beneficial to your medicinal cannabis. The smaller leaves that are covered with resin can also be cut away but can be used in smoked cannabis or for hashish.

Drying and Curing

To begin drying, cut off the trimmed branches on the individual buds from the plant. In a well ventilated, cool, dark place hang these pieces upside down. Make sure there is space between each bud, they should not be touching.

The drying process should take between four to ten days. Check the stems of the buds. When the smaller stems are fully dried, they will snap when you bend them, the thicker ones shouldn't. Now you're ready to start actively curing the buds.

Curing is meant to enhance the quality of the buds before you smoke them. The curing process begins when you stop watering your plant, so while it is being dried, it has also begun to cure. To continue the curing process, you'll need jars and lids. Fill these jars to what looks like 80% full and tighten the lids. Store the jars in a cool, dark place.

For the next two weeks or so, open the jars once a day for a few seconds to get fresh air into the jars and to let any built up moisture evaporate. Similar to the drying process, you'll need to watch your jars and levels of humidity so mold doesn't begin to grow. If the buds feel wet at any point you can open the tops of the jars and allow the buds to be exposed to air until they're dry on the outside. After a week of the buds feeling dry to the touch, you can open the jars once a week instead of every day.

A properly dried and cured bud will not have the grassy smell any longer and will result in cannabis that does not have a harsh taste.

How to Increase Your Yield

Growing marijuana is an enjoyable experience; after all you're growing something to benefit yourself and other people suffering from illnesses. But, you are not growing this for the fun of it, the bigger the yield from your plants the happier you and your will be.

This means you need to know how to increase your yield and make the most of the plants you have. These are the techniques which will make this possible:

Light Intensity
As already discussed, maximizing the amount of light your plant gets during the growth stages will help to ensure that it grows large and provides plenty of buds. If your plants are tall and skinny, they are not getting enough light. Try leaving the lights on for longer and bringing them closer to the plants.

Plant Training
While you want your plants to stretch outwards you also need them to grow tall and to only have large buds. This might sound like a tall order but with a little understanding you can train your plant to grow right. The first step in achieving this is not overcrowding the plants when they are little. Each one needs to have the room to stretch out without conflicting with other plants. If they overlap they'll receive less light and heat, resulting in a poor yield.

Your baby plants should have approximately 3 inches (8cm) all round their main stem. Providing they have this space pack in as many as you can, the more plants the bigger the yield! If you have room for 4 large plants you should seed 10. If the room becomes too small you should get rid of the weakest ones and continue with the strong ones.

Bend Your Plant

In order to maximize the light to each flower, you can bend the stems horizontally. You must always bend them near the bottom to avoid breaking them. This will allow more light to the buds and encourage more colas to grow; increasing your potential yield. It is worth noting that if you partially break a stem when bending you can wrap it in tape and it will heal itself.

Bud Arrangement

Once the plant is growing you need to keep an eye on any holes in its canopy. You should be able to bend other colas into these places to ensure that every part of its surface has buds, maximizing the yield from each of your plants.

Crop It

Once you clearly have predominant stems and a good coverage across the plant, remove all he stems that are hidden. They won't produce buds but they will deprive your healthy buds of nutrients. Any bud that's not in direct light can be got rid of; be ruthless!

Feminization

It is extremely difficult if not impossible to determine whether your plants are male or female when they are young. Male plants exist to pollinate the female plant; this means that they are not interested in providing you with nice buds or increasing your yield. Worse, if they succeed in pollinating the female plant the focus will shift from producing buds to seed growth; your yield will decrease!

However, because you can't tell a female seed from a male seed you need to ensure you are planting female plants. This process is known as feminization and is an important step towards increasing your yield.

Of course you can simply purchase female seeds but this allows you to complete the process yourself:

Silver Thiosulfate - This is the most common approach. You need a solution of silver nitrate and sodium thiosulfate. You will need to choose one female plant that is nearly mature and then add the solution to its soil. As the solution is absorbed it will trigger a stress response the plant which will change the gender of male seeds to female ones. Which will result in giving you female seeds for your next batch.

Colloidal Silver - An alternative is to add some colloidal silver to a spray bottle of water. You can then spray your female plants when they start flowering. This will create female pollen sacs and ultimately female seeds.

Rodelization - The third option is to stress your female plants; this will cause the same hormone change as Silver Thiosulfate, turning your male seeds female. This is the least popular or effective method. You simply make sure that your flowers stay for at least 3 months, by keeping the light right. Then the plant will need to opt for survival by creating female seeds.
The key to this approach working is to make sure your plant is truly stressed. You can do this by using a 12 hour light then 12 hour dark cycle for one week, then giving her 24 hours of light before going back to 12 on and 12 off. The stress will cause the hormone changes which will create female seeds.

Storing Your Seeds

As already mentioned it is not possible to distinguish between a male seed and a female seed. Therefore, once you know that you have female seeds you need to keep them together. The best way to store the seeds is in a paper bag; this will absorb excess moisture protecting your seeds.

Do not store them in an airtight container, the moisture in the seeds will create mold and ruin them before you have a chance to plant them.

If you want to start, the seeds can be placed onto a damp piece of kitchen paper and left for 24 hours in a dark place. This will allow them to produce their tap root which will enable you to plant them in a suitable size pot and start growing your next batch.

The female plant will have male and female pods growing on it. But, as soon as the plant reaches the pre-flowering stage and the pods are visible you need to look at each one. Any pod which has a little green sac under it, where it joins the stem, is male. You need to remove them before the sac bursts open. If it does your yield will be ruined. But if you successfully remove all the male buds and sacs then you'll have a good yield and a potent plant.

Genetic Selection

It should already be obvious that there are a huge number of different strains of marijuana. In fact, more and more are being produced every year. The result can be overwhelming and can leave you questioning whether the plant you have is actually the one you think you have or that you wanted!
This is why it is essential to understand the genetics of marijuana and the best methods of selecting your plants.

Genetic Stability
You might have heard of this term or heard someone mention that a specific plant is not stable. There are actually two factors which contribute to the genetic stability of a plant; these are discussed in a moment. For the sake of clarity, it is useful to understand that 'genotype' refers to the DNA of a plant while 'phenotype' is the identifying factors of a specific genotype.
In a nutshell, the genotype of your plant can be influenced by the environment you create for it. The result is a specific phenotype.

You can liken it to child raising, their genes plus their experiences as they grow makes them into the person they are today.

1. **Variability**

The result of genes and environmental factors is more varied the less control you have over the environment. For instance, plants grown in the wild will flower when nature tells them and cross pollinate according to what plants surround them and the direction of the wind.

In order to produce a plant that has your desired characteristics you'll need to control the environmental factor as closely as possible. This is done by using this book, your enclosed growing room and by regulating the light, temperature, humidity and nutrients. The result leads to the predictability factor.

2. **Predictability**

There are specific genes in plants which cause them to have purple flowers or thin pointed leaves. Knowing the seed you have will allow you to predict the type of marijuana you will end up with. This is the strain.

Providing you can control the variability of environmental factors then you should have a plant that produces the right type of marijuana. But, if you get the environment wrong then your plant can be influenced to change its flavor and be more variable.

A stable genetic plant behaves exactly as you expect it to and this is achieved through a mixture of controlling the environment and choosing the right strain.

Using Your Gene Knowledge

Now you know how the genes and the environment affect your plants you need to consider the best way of cultivating the characteristics you desire in your plants. To do this it is important to understand the power of these genes.

To create a new plant, you'll need a male and a female plant. When they pollinate the new seed will contain 50% of the gene code from each of its parents.

For example, if a female plant has genes AA then the new plant will take one of the A's. However, if her genetic code is AB then the new plant could take A or B; you won't know which. You need to have plants that you know both gene sets are the same, this will allow you to reliably predict the characteristics of their offspring.

How to get the characteristics you want

Unfortunately, there is no shortcut for this process. You'll need to have a female with known characteristics and a male. You can then isolate them to ensure they breed together and look at the results.

You'll need to wait for the growing cycle to be complete before you can identify the characteristics of your new plants. You'll need to isolate these plants and allow them to fertilize again. Once they have grown you'll be able to check and select the best two to continue narrowing the gene pool. It will usually take several generations of plants until you can be confident that both the male and female plants have the same genes and will predictably produce the characteristics you want.

You can then use these plants to produce as many seeds as possible and these will become your marijuana plant of the future.

They will provide guaranteed characteristics but you must keep them isolated from any other types of marijuana, or you'll have to start the process all over again. If you buy seeds online, it's already take care of and you don't have to worry about this.

Jump starting the process

It is possible to jump start the process by purchasing female seeds which have been cultivated by a professional. This will guarantee you get the characteristics you want. However, all these seeds will have been feminized so you will struggle to produce additional plants in the future. Completing the gene pool process yourself allows you to create as many plants as you want in the future. More importantly you'll find that it can allow you to experiment and discover new strains.

How to create a new strain

A new strain is created by allowing a male of one strain to fertilize a female of a different strain. You'll need to adopt the above approach to genes in order to develop a strain that is predictable every time. This is how the best and most potent strains have been developed. You might find it of interest to understand these terms:

Landrace – These are plants which have been cultivated in the wild or allowed to grow without interference. It is very difficult to ascertain what phenotype they will have any how effective they will be as a medicinal aid. These plants are deliberately varied as this increases their chances of surviving in the wild.

Cultivar - This refers to plants that have been developed and cultivated by humans to create specific characteristics. These are predictable and cover the majority of the plants on the market today.

Selecting the right plants

The first thing to consider is that only the strong survive in the wild. You need to act like mother nature to get rid of any weak male plants; ensuring only the best are able to cultivate your chosen females.

1. push down on the middle of a branch with your finger. Don't overdo it, medium pressure is enough. If the stem breaks ditch the plant.

2. If you have found a strong one, look at the leaf structure of the plant. Lots of high leaf cover will block the growth of the plant. This is not a good specimen to use. The less maintenance you need to do the better.

3. Check the scent of the male plants in the vegetative state by rubbing the stems. A strong smell is a good sign that you have a strong male.

4. The flowers should be dense and healthy. The larger the cluster of flowers the better. You can also assess the amount of pollen which they give off when they open, although most plants have a similar quantity.

Once you have selected a strong male you need to pair it with a good quality female. Then hope you have the mix right. Of course, before you start this process it is essential to understand the characteristics of the strain you have. You can't pass down characteristics which don't exist!

Initial Stock

This is perhaps the most important part of all. Your initial seeds will need to come from somewhere and that's probably an established breeder. Before you select your seeds you need to know the strain you want, you can do this by looking at the different strains discussed in this book.

Then, you'll need to research your potential breeder. You need to find one that has a good reputation. Unfortunately, the breeding industry is unregulated and there are plenty of people who will sell you anything. To get the best genes and shorten the time it takes you to get predictable yields you must use the best quality seeds. That means using the breeders with the best reputation. There are several seed banks per state.

It would be a very long list to mention them all here. You can do your own research on google by using this search phrase:

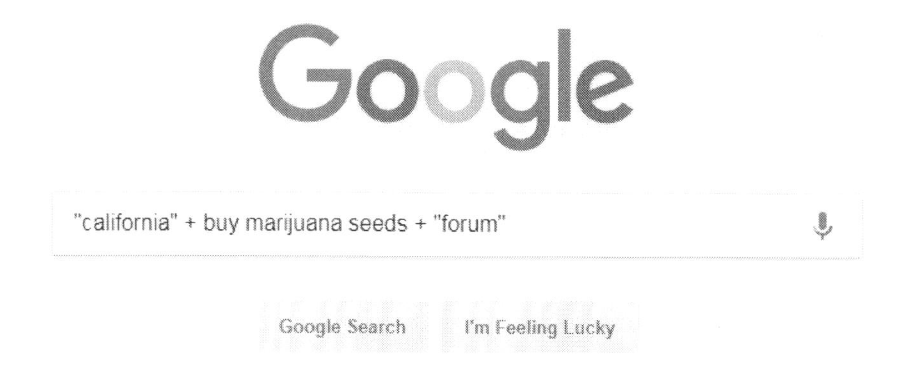

If you use this phrase you will find results on forums where people buy seeds from online.

Indoor vs Outdoor Growing

Indoors

Growing indoors is a great option if you're tight on space, don't have access to fertile land, or don't live in a particularly sunny environment. To get started you'll need a space like a spare room, closet, garage, attic, garden shed, or bathroom. Whichever space you choose, you must have a window or a vent where you can get rid of used air and electricity so you can plug in your lights. Alternatively, you can buy or make a space specifically designed to grow your plants in, like a grow tent or grow bucket.

Grow Tent or Grow Bucket

A grow tent is a lot like what it sounds like. It's lightproof, has reflective walls, a waterproof floor, and there are places to hang lights and vent your tent. A grow tent also features an opening at the bottom for fresh air to enter and an opening at the top for used air to exit. There are also openings for electrical cables to feed through.

A grow bucket or space bucket is a planting tool you can make or purchase that consolidates your grow medium, plant, and light into one space efficient bucket. You can shop for premade versions of either of these row spaces or build your own.

Outdoors

To grow outdoors, you'll need a very specific environment. Not only does that mean space and good soil, but you need to live in an area where you get at least eight hours of direct sunlight. Growing outdoors also means you're limited to growing with seasons: planting in spring and harvesting in fall as opposed to growing year-round with artificial lights and timers.

This is the cheaper option if you can swing it but you also run different risks than if you grow indoors. There's the issue of privacy from neighbors and people potentially stealing your plants as well as the environmental risks of animals, bugs, and bad weather. All of these factors can have adverse effects on your plant's water intake and temperature exposure.

It's best to start planting your seeds in the spring. In Australia and South Africa for example, Spring is in October. In Germany and The USA it is late March.

Cost

Generally, it costs less to grow outdoors. You're not paying for electricity or to moderate the humidity- just soil, seeds, and nutrients.

Growing inside will cost you more but it also depends on how ambitious you are, how much you want to grow, and which lighting and grow medium you've chosen. DIY is a strong value in the world of homegrown cannabis. You can buy a space bucket just as well as make your own, clone your plants instead of continually buying seeds.

So if you're a medicinal cannabis user or caregiver on the East Coast or in the Pacific Northwest where the weather isn't great but you also don't want to spend thousands of dollars, you might want to consider having a DIY setup.

Troubleshooting Guide

Pests

Although the risks of pests getting onto your plants is higher if you are growing outdoors, there are still plenty of bugs which are eager to damage or even destroy your hard work.

Unfortunately, if a disease does attack your plants when you're using a growing room, it can quickly spread and do a lot of damage. In the wild you can help to prevent disease from spreading between plants by spacing them out. Of course this is not an option in a growing room, space is at a premium.

By being aware of the most common pests and knowing how to deal with them, you can prevent your plants being attacked by red spider mites, thrips and even white fly.

Know Your Soil

If the soil you have been using is not sterilized, you run the risk of having eggs in the soil. As the weather warms, or your growing room reaches optimal temperatures the eggs will hatch leaving you with a pest problem. Having an indoor crop means that these bugs will have unrestricted access to your crop; it's unlikely that a natural predator will also happen to be in your growing room.

Because the growing room is set for maximum growth any bug will be free to multiply, your entire crop could be damaged in a matter of days. Paying for sterilized soil rather than using soil from the garden is an exceptionally good use of your money.

Watch Your Hygiene

Because your plants can be extremely vulnerable, you need to ensure the right preventative methods are in place to stop any pest arriving on your plants. This means you need to take hygiene very seriously.

The first step is to always wash your hands or wear gloves when handling your plants; bacteria are everywhere. A recent survey showed that there can be as many as 10 million just between the keys of your phone!

Temperature

You already know that the temperature needs to be kept right to ensure the best possible growing conditions. However, it is also essential to avoiding pests breeding amongst your plants.

But temperature does not work alone; it needs moisture to attract the pests to your plants. This means that during the early stages of cultivation the conditions in your growing room will be very attractive to pests. Keeping the temperature and humidity levels right is a step forward in the battle against pests.

Entrances

The biggest risk is the bugs and bacteria that you can bring into the room. Avoiding the room is not an option although keeping pests to a minimum is a good reason to only enter the room when you really need to.

Preventing bugs from getting in means checking the potential entrance points. Any ventilation system you have which leads outside must have anti-thrips nets incorporated into the pipes. These are fine enough to stop the majority of pests from getting into your room.

You will usually only have one other point of entry, this will be the doorway. By making sure you are clean and leaving your pets outside.

Select The Strain

There are several strains of marijuana plants that are more resistant to pest. These include Easy Haze and Heaven's Fruit. Choosing to plant these will help to protect your plants; providing this is a strain you want!

Know Your Pest

I've mentioned the red spider mite, thrips and white fly. If you're going to grow marijuana, it's important that you know what these pests look like.

Red Spider Mites - This pest is so small it is incredibly difficult to see yet it can cause a huge amount of damage to your plants in a very short space of time; they will suck the enzymes from your plant. This will reduce the brightness of its color, the potency of its product and prevent it from absorbing enough nutrients!

You'll need a magnifying glass to look very carefully on the underside of your plant leaves. You're looking for any irregularities. It's unlikely that you'll see the actual mites. But, you should see the black dots which show it's there.
These black spots are the eggs. Look closely and you'll also see a web type residue sticking to your plants; this shows the red spider mites have been there. If you find these warning signs then act fast, before your crop is ruined.

White Fly - These little pests also feed on the underside of the leaves. However, you won't need to inspect your plants to spot this pest.

If you see them flying round over the top of your plants then you are probably looking at white fly. The flies are small but can be seen by the naked eye. If you turn a leaf over you may see them on it but it is more likely that they will be disturbed as soon as you walk near the plant; hence the flies over the plant. Again, prompt and decisive action is essential.

Thrips - Unsurprisingly thrips are attracted to the underside of your leaves. The adult thrips are straw colored or black, and no bigger than 2cm or ½ inch.

Thrips are known to spread an array of diseases. Your plants will start to turn pale; they may display an abundance of spots and then take on a silvery look before dying.

Dealing with these pests

If pests appear despite your efforts to prevent them then you need to be ready to react. The problem is that you can't use conventional pesticides. They will damage the flavor of your plants and could even create a dangerous strain. You will find these natural alternatives effective:

Cinnamon Extract - This extract is very effective at killing various mites. It can be safely sprayed onto your plants throughout the growing cycle but you should stop 10 days before you're ready to harvest your crop.

Potassium Soap - This soap kills the insect as soon as it touches them. You can purchase it in most general stores and it is particularly effective against white fly and thrips. Apply liberally every day to destroy the infestations.

Neem Oil - This oil is taken from the neem tree and is excellent against spider mites, white fly and thrips. The best way to introduce this is through the water system, whether you water them manually or with an irrigation system. It kills as soon as it touches a pest. Again, it is not advisable to use this if you are 15 days or less away from harvesting.

Propolis

This is actually made by bees; it seals their hives. But, it also has excellent anti-fungal properties. Mixing a little with the soil of each plant will help to prevent mold and other bacteria forming.

Make Your Own - Garlic, vinegar and even onions are all sad to be effective at either keeping pests away or eliminating them. If you are feeling brave you can create a mixture from 2 tablespoons of the juice of each of these ingredients. Spray it on your plants to kill the pests.

It is worth noting that if you do develop a pest problem you can, if possible, isolate the affected plants. This can help to prevent it from spreading.

Seedling Issues

Transfer Timing

Do not transfer seedlings until after your plants have grown the initial set of leaves. Transplanting to early can cause stress and damage to the plant.

Nutrient Amounts

In terms of nutrients, you don't want to rush this either. If you're planting in a soilless medium or through hydroponics, only add nutrients at seedling strength, which is about ¼ of regular strength.

In the very beginning, before your seedling has many leaves, you may leave a CFL light 6in away from the plant as long as it passes the hand test. After its first two sets of leaves have developed, move the lights as close as 2in as long as it passes the hand test.

When in doubt, check the pH of the soil to troubleshoot any perceived seedling issues.

Nitrogen Toxicity

If the leaves on your plants start to bend at the ends and look like a claw, if the leaves are a dark green, if the stems are weak, and if the overall growth is slow, you're likely looking at a case of nitrogen toxicity.

If left untreated, the leaves will turn yellow and the plant will die. It's a common mistake to give a plant too much nitrogen, especially in the flowering stage. This is where the danger of using a time release soil comes in. With a time release soil, the plant will continue to get the right amounts of nitrogen throughout its life instead of varying levels as it matures. Here's the nitrogen plan you should have your plants on:

Vegetative Stage: High levels are good here. Any complete plant food will work just fine.

Flowering Stage: Drop down to lower levels here, mixes labeled "Bloom" or Cactus nutrients.
You can't reverse nitrogen toxicity, but you can prevent new leaves from being affected.

Calcium Deficiency

Calcium deficiencies tend to show up on new leaves, so you want to check your new leaves for any:

- Dark green leaves with dead or brown spots
- Crinkling
- Curled tips
- Stunted growth
- Weak or hollow stems
- Underdeveloped, weak roots
- Slow or undeveloped buds
- Decay

Calcium helps the plant withstand stress and provides structure for the plant. You likely won't be overdoing it on the calcium if you are using normal amounts of nutrients or regular soil.

Calcium deficiency is likely to occur:

When growing outdoors: because the soil can be more acidic, which is a pH below 6.2.

When using soil: and you haven't added a calcium supplement like dolomite lime or the water is acidic. When growing in soil, calcium is best absorbed by the roots in the range of 6.2- 7.0 pH.

When using coco coir: and you haven't added a calcium supplement or the water is acidic.

When using a hydroponic system: and you've neglected to use nutrients that will supplement the calcium or growing in water that is acidic. When growing in a hydroponic system, calcium is best absorbed by the roots at a pH between 6.2- 6.5.

You'll notice that acidic water and soil are the most common denominators. This is because when the pH is off, the plant cannot absorb calcium through the roots. Therefore, a good first step in figuring out if your problem is calcium deficiency, you should check the pH levels. Your second step should be to flush the system with clean, pH water with normal levels of nutrients.

After you've taken action against calcium deficiency, pay attention to the newly growing leaves. The brown and dead spots already on the leaves will not go away, but new leaves should be coming in healthy and strong within a week.

Common mistakes to avoid

Whether you are attempting to grow marijuana for the first time, or if you have years of experience, you'll find it useful to now the most common mistakes made. This will ensure you don't make the same mistakes and increase your chances of having a successful crop.

Not Being Observant

Pests are an issue as we have just discussed but they can be spotted and dealt with. However, this means that you need to be vigilant. Walking into your growing room and turning over a couple of leaves is not enough. Although checking every single leaf of a big crop is not possible, you do need to check a good quantity of them and do so every day. This, along with paying attention to hygiene can help to ensure your crop is successful.

Lighting

By now you should be aware of the importance of light to control the growth of the plants and even start them flowering. The problem is in balancing the amount of light each plant gets without over heating them or having the lamps close that it actually burns their leaves.

It is impossible to state a set distance that is acceptable, there are too many variables involved. The power of the lamp, type of lamp and even the size of the growing area all need to be considered.

As a basic rule of thumb you should be able to hold your hand just above the leaves without feeling like its burning. You can even use a thermometer to check the temperature at the top of the plants. It should be 28 – 29° C (82 – 84° F). But don't forget, the closer the light is to the plant the fewer the number of plants it can look after.

Spraying

A common problem, mold, is cause by the conditions being too moist. This attracts the bacteria which create mold and allow it to spread rapidly. Most growers realize the importance of controlling temperature and humidity but many still use a spray option to water their plants. Spraying the leaves will dramatically increase the chances of mold forming, which will damage your harvest.

Leave Your Plants Alone!

Once you have set your feeding, watering, light and pest control schedules there is no reason to disturb your plants. Just like human's plants can become stressed and this will have a detrimental effect on their growth and your harvest. Your plants are capable of growing well with the measures you have put in place. After all, cannabis is still a weed.

Panicking

People who are new to growing marijuana can be panicked by the sight of dying leaves and start creating radical changes to their systems. In fact, it is normal for low level leaves and those on the outskirts of the light to change color. They can simply be removed, leaving the plant to flourish. Instead of panicking when something doesn't appear right, pause, take a look at this book and then take the appropriate action; if any is needed.

The Root

The fundamental element of a productive and healthy plant is a well-developed root ball. This is a good reason why you should consider re-potting your plant regularly. You will be able to verify that the roots are starting to grow through the bottom of the pot, signifying that the roots are growing well. A healthy and large root ball is essential to ensure the flowering buds all have the nutrients they need.

Low Level Cold

If you've fitted a thermometer to the top if the plants and established that they are just the right temperature you have only completed half the battle. Heat rises, this means that the floor can have a much lower temperature than the ceiling. Your plants might be warm at their tops but have cold roots; leading to a slowing of nutrient uptake and potential damage to the plant! Monitor the temperature at the bottom of the room and consider adapting your heating system to ensure the temperature is even across the room. Additional low level heaters and fans can help to ensure the temperature is consistent from the top to the bottom of your plant.

Rushing It

If you push your plants to flower too early then you'll have less of a yield that you hoped for. You will also notice it is not as potent as it could be! This is a common rookie mistake and one that you are not likely to make twice.

But, one thing that experienced growers do is attempt to dry the harvest quickly. The quicker it dries the earlier you will be able to use it.

However, if you use artificial means, such as a hot air blower, to dry your harvest you will increase the levels of chlorophyll. This will destroy the potency and flavor of your harvest. In addition, if you rush to wrap your product then you are likely to do it before it has fully dried out. If there is more than 20% humidity in it when it is wrapped, then you'll experience mold and bacteria growth.

Starting Too Big

It should be obvious that there are many factors to consider when growing marijuana. No one can get it all right on the first attempt; even if you buy every piece of state of the art equipment available. High quality equipment can only be used in conjunction with knowledge. This is why it is better to start small, learn and build your setup as your growing skills improve. Trial and error makes a better grower and is how many of the strains were discovered!

Listening to Everyone

Everyone has an opinion on growing marijuana, whether they grow it, use it or don't even know what it looks like. This is a problem. Even the most advanced growers can only provide general advice on your set-up; because every setup is different and you'll need to learn what works for you, your space and the strain you are growing.

If you try listening to everyone you'll get conflicting advice and probably give-up when your new hobby could be so much more. Use this book and don't be afraid to experiment; it will make you a better grower.

Listening to The Right People

The opposite is also true! While no one can tell you exactly what will work for your growing room, they can provide advice based on their own experience. They can also give you insights into what should work and how you can improve your yield. Using the appropriate resources; this book and people with large amounts of experience, can help you to find the right balance for your plants. You just need to be selective about the advice you accept.

Check Your Power

An average house uses 20 amp circuits in each room. One breaker (110v mains) can handle a total of 2200watts. Mostly there is one circuit breaker for each room. Check your electrical plans if you need more power and run an extension cord from one room to the other in order to use more wattage.

Grow Room on a Budget

As you can now see there are many different elements involved in the production of marijuana. It can be very beneficial to use the following case study as a guide to getting started.

This example assumes a starting budget of $350; of course, if you had more available you could upgrade the set up and increase your yield.

It is also worth noting that many of these products can be purchased pre-owned. This will reduce the cost of your set-up but you should verify each item is working properly before you part with your hard earned cash.

General Hydroponics FloraGro 1 Quart - $13

Having the right nutrition for your plants is essential. This product is specifically designed to help build strong roots in any plant. It also encourages vegetative growth.

This makes it perfect for growing marijuana as your emphasis must be on the vegetative state. Once they start flowering your job is nearly over!

This nutritious product is so good that NASA and Antarctic research scientists also use it. It has all the important nutrients, phosphorous, nitrogen, potassium. It is suitable for traditional soli based plants and those growing hydroponically.

You'll need to add approximately 2 teaspoons of this to a gallon of water. In most set-ups once a week should be sufficient but you'll need to assess your plants yourself to confirm what works best.

One bottle should be sufficient for at least 10 plants for their lifetime. This makes it good value for money!

Amagabeli 6 Inch Carbon Filter - $62

I've mentioned the importance of keeping the odor contained. The better your plants are doing the more pungent the odor will be. This makes it imperative that you choose a high quality carbon filter set up. This is one area where you cannot afford to cut corners.

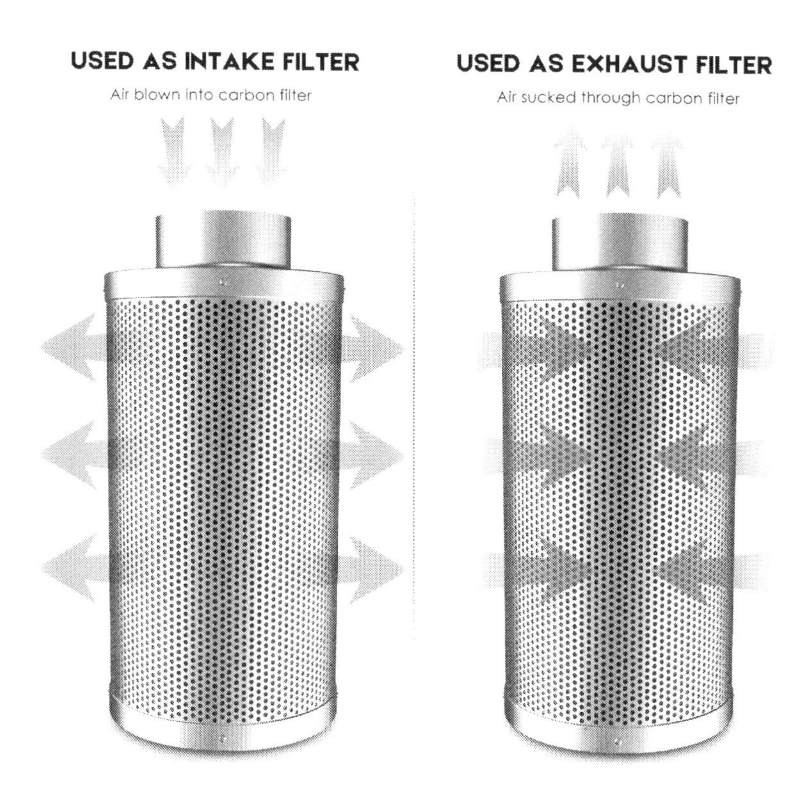

This filter is made from aluminum. This means it is light and can be easily moved if you need to relocate your growing room for any reason. It has a packed carbon base which will ensure that 99.8% of odors are captured and eliminated; that's impressive.

Because it is 6 inches it will fit inline in the most commonly used ducting pipes and inline fans. In addition, it has an inner and outer mesh which increases the amount of air it can deal with on a daily basis.

The main aim of your set-up is to remove odors but you may also need to control the air coming into your room. Fortunately, this filter can be used as an extractor and an intake fan.
This is why it makes the perfect addition to your set up. You can switch it between intake and extract by simply changing the direction of your fan; allowing you control over the air supply in and out of your grow room.

But, most importantly is how easy it is to use. You can attach it to ducting and an extractor fan to create a high quality set-up.

6 Inch Aluminum Foil Duct Hose Flex - $16

To complete your set-up properly you need ducting. This has to be 6 inch in order to accommodate the filter listed before.

The reason that this is one of the best choices is that it comes with the clamps to secure it to the fan and any other piece of ducting you need. It is essential to have airtight seals or you'll ruin the effectiveness of your entire set-up.

But, more importantly, this pipe is also aluminum. It is lightweight and will not place a burden on the ceiling of your grow room; especially if you use the same grow tent as I recommend.

It is designed as a cooling duct but it is also very effective as heating duct. This is important as your grow room will become warm and you don't want your ventilation system getting heat damage.

It is flexible and fits perfectly with the filter above and the fan below, means that you'll have an airtight ventilation system which can be weaved round your plants to create the maximum effect.

Hydroplanet 4 Inch Duct Booster Fan - $20

You probably think that virtually any fan will do the job. After all, all you're trying to do is move the air in and out of your grow room.

But there is a good reason why this is the best choice you can make. It is high powered which will ensure it generates enough air movement to keep at least 5 or even 20 plants happy. This is achieved through the power of the motor and the duct boosting fan.

The fact that the ducting is slightly larger ensures that the fan is never straining to pull in air; ensuring it does the best possible job. You'll also find locking tabs incorporated into the fan; ensuring it easily and securely connects to your ducting.

You can also add a controller into the system to tell the fan when to switch on and off. It will also fit the vent hole of the grow tent perfectly! Of course the fact that it works very quietly is certain to be beneficial.

Calipots 5-Pack 3 Gallon - $20

You need pots to grow your marijuana. While you can purchase pre-used ones it will be important to clean them and sterilize them thoroughly. This will ensure you are not bringing bacteria and other contaminants into your plants.
However, the time and cost of doing this means it actually makes more sense simply to purchase these high quality pots which are ready to use.

They are injection molded with heavy rims to make it easy to move them; if you need to.

They are designed with drainage holes. Assuming you use the soil recommended in this guide won't have any problems with rotting roots. The top of the pots are 11 inches, allowing you to comfortably place 5 inside the grow tent below.

Vivosun 48"x24"x60" Mylar Hydroponic Grow Tent - $70

You can create a grow room as described in this guide. But, particularly for your first attempt, you'll probably find it easier to purchase a grow tent. This will ensure the space is completely dark when needed.

This tent is designed with reflective material inside to help maximize your light. It has an observation window to allow you to check on the plants without disturbing them. Of course you shouldn't open the flap for the observation window during their dark cycle. The entire tent is double stitched to ensure complete darkness inside.

As a bonus it has a vent hole ready for your fan and ducting. This means you can have your set-up created in virtually no time; allowing you to concentrate on growing plants.
This is not just for people trying to grow marijuana for the first time. The Mylar floor tray makes it great for experimenting with a hydroponics set-up. This will allow you to try out new techniques without disturbing your main grow room.

Apollo Horticulture 600 Watt Grow Light - $160
Of course, no set-up is complete without the light! There are many options available on the market but this is one of the best. One of the main reasons for this is that the light is dimmable. As your plants grow their light requirements change (as already discussed in this guide).

Combining this with your grow tent and its reflective walls will ensure you have plenty of light for the 5 - 10 plants you can comfortably fit in there.

It is worth noting that this light can be dimmed to 50% or 75% of its normal power. It has 1 HPS bulb and 1 MH and comes with a set of rope hangers to facilitate your installation.
This really does cover all the options throughout the growing cycle; making it the perfect addition to your set-up.

Shopping Cart

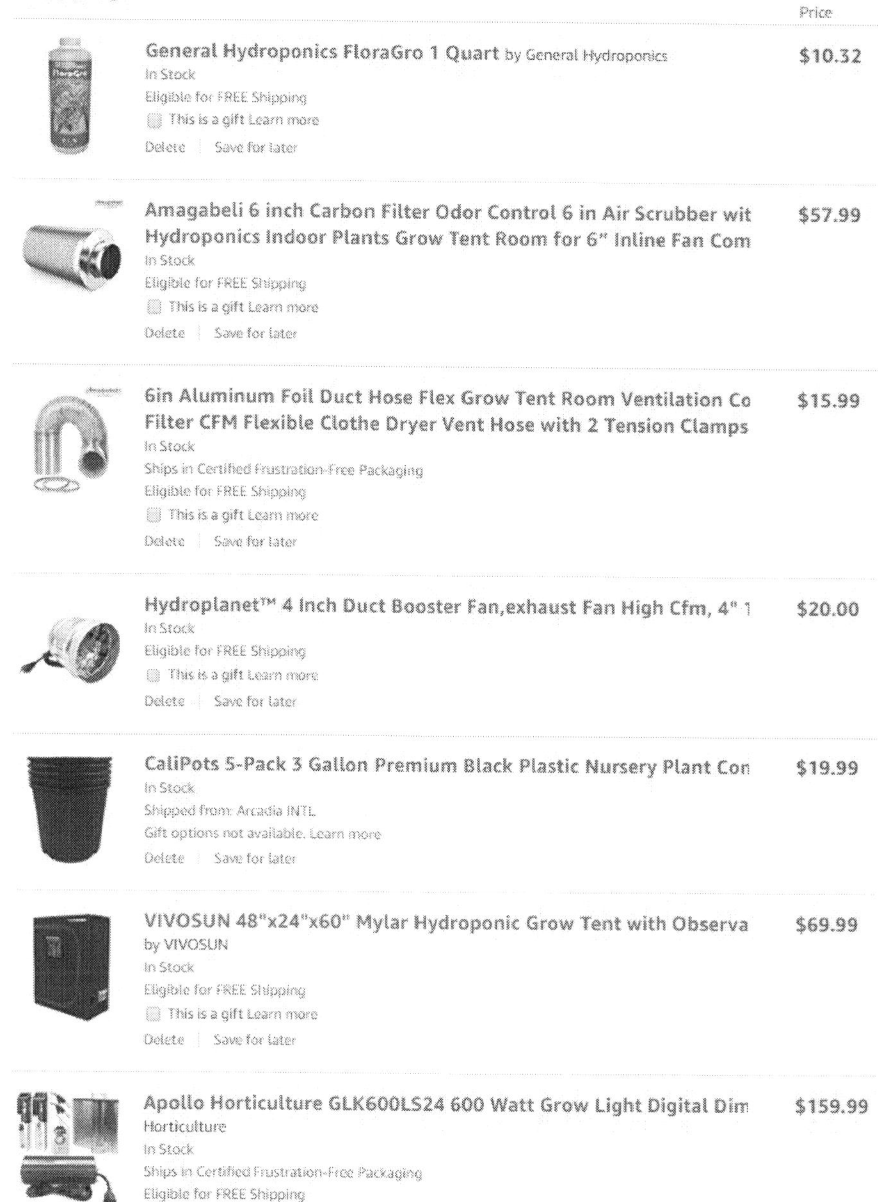

Summing Up

You've now got everything you need to get started. It is recommended that you complete the process in this order:

- Create the ventilation system by making holes in your room, adding the filters, fan and ducting. This will bring fresh air in and remove warm air.

- Black out the room, don't forget to test this several times to ensure the room is pitch black. Even a fraction of light can prevent the flowering stage from starting.

- As part of this process you should add your reflective Mylar to the walls. This will help to keep light out and it's better to do it before you check the darkness levels. If you just complete the blackout part, you'll only have to check again after installing the Mylar to ensure you haven't damaged the black out!

- Add the flooring to help ensure the moisture levels are low and no damage is done to the original flooring. (If this is an issue).

- Position your plant pots. It is best to create a simple wooden structure which will lift the pots above the ground. Not only will this make it easier for you to look after them it is easier to keep the floor clean. This is important to minimize bacterial activity.

- Install your lights, making sure that they are above the area where you want to grow your plants and below the ventilation ducts.

- You will need to make sure you can adjust the height of these, a simple chain or an additional block above the fitting is all this takes.

- Add your heater at ground level to ensure the lower part of the plants stay as warm as the top. You will need to use a heater with built in thermostat to make this efficient.

It is important at this point to make sure you have the light set up right, you want your plants to flourish in the vegetative state before you allow them to flower.

As well as light, watering and nutrients are important; as already described in this book.

Optional Extras

Moving up from a budget installation there are a number of extra features you can add which will help your crop to have an even better yield. This will increase your budget or you can look at ways to reduce your initial expenditure by purchasing pre-owned equipment. Providing it works there is no reason why this is not a viable option.

Optional extras include:

Timer switch - This will allow you to automatically switch on and off your fan depending upon the heat or the moisture in the air. There are plenty of these on the market which are designed to connect with your fan and are easily wired in.

Humidity Meter - This is a useful addition, especially as you move into advanced growing techniques. It will enable you to track the humidity of your grow room and perform small changes, while knowing how they affect the humidity. The consequence will be a better yield.

EC Meter - The best way of administering the nutrients is by adding them to a large bucket of water and allowing them to sit for several days.

An EC meter will measure the nutrients in the water to ensure you are giving the plants the right amount every time!
As with any type of gardening your success will be dependent on your level of commitment and the tools at your disposal. Controlling the light and temperature are the most important parts of this process. But, ensuring nutrients, watering, pruning and even checking the buds can all make a huge difference to your yield.
The most important thing to remember is that this is a learning experience. The more you document what you have done the easier you will find it to adjust it and improve it for future yields.

Growing marijuana should never be considered a one off thing, it takes a small budget and time to perfect the process, but then the results are more than worthwhile!

Conclusion

Growing marijuana is not something you start and then give up a few months later. If you are going to commit to this then you need to consider all the options and create the right setup before you purchase your first seeds.

There are many different factors to consider, from the light levels, to temperature, humidity and even which system you are going to use to feed your plants.

The best way to become an established and experienced grower is to start small and slowly make changes to your systems to create the best possible environment for your plants.

Marijuana can be a potent medicinal aid, growing it yourself can save you a large amount of money and it is extremely satisfying. However, as this book has told you, it is important not to rush the process.

You also need to be aware of the laws in operation where you live; you don't want to fall foul of them. It is vital to remember that even though you are growing within the law; your ventilation system will send the distinctive aroma of your plants into the open air. This can attract undesirable attention, from concerned neighbors to others who want your product and don't want to ask you or pay you for it. Save yourself this hassle by implementing the measures described in this book. Carbon filters are cheap and extremely effective.

The better your understanding of each of the growing stages and the lighting requirements, the easier you will find it to grow high quality marijuana. There is no doubt that the growing environment is easiest to control when you have an interior growing room but this guide can be effective at helping you to grow outside; if this is your preference or only option.

Creating stains for yourself is a long term commitment. It can take generations to obtain the strain you want. This time period will also allow you to develop the systems you need to support and nurture your plants with the least amount of intrusion.

The result will be quality marijuana and a sense of satisfaction that you have managed to grow it successfully and consistently. The tips in this book will ensure that you can go from beginner to advanced, with as few issues as possible. But don't forget that growing marijuana is a learning process. There are always new techniques to learn and ways to improve the quality or quantity of your crop.

Lastly, I would like to thank you for getting this book. It has been a pleasure to write it.

I would like you to ask a favor. If you have found this book helpful, could you please leave a review for this book? That would be awesome!

Thank you and happy growing!

Notes